Propositional, Probabilistic and Evidential Reasoning

Studies in Fuzziness and Soft Computing

Editor-in-chief
Prof. Janusz Kacprzyk
Systems Research Institute
Polish Academy of Sciences
ul. Newelska 6
01-447 Warsaw, Poland
E-mail: kacprzyk@ibspan.waw.pl
http://www.springer.de/cgi-bin/search_book.pl?series=2941

Further volumes of this series can be found at our homepage.

Vol. 55. J.N. Mordeson, D.S. Malik and S.-C. Cheng
Fuzzy Mathematics in Medicine, 2000
ISBN 3-7908-1325-7

Vol. 56. L. Polkowski, S. Tsumoto and T.Y. Lin (Eds.)
Rough Set Methods and Applications, 2000
ISBN 3-7908-1328-1

Vol. 57. V. Novák and I. Perfilieva (Eds.)
Discovering the World with Fuzzy Logic, 2001
ISBN 3-7908-1330-3

Vol. 58. D.S. Malik and J.N. Mordeson
Fuzzy Discrete Structures, 2000
ISBN 3-7908-1335-4

Vol. 59. T. Furuhashi, S. Tano and H.-A. Jacobsen (Eds.)
Deep Fusion of Computational and Symbolic Processing, 2001
ISBN 3-7908-1339-7

Vol. 60. K.J. Cios (Ed.)
Medical Data Mining and Knowledge Discovery, 2001
ISBN 3-7908-1340-0

Vol. 61. D. Driankov and A. Saffiotti (Eds.)
Fuzzy Logic Techniques for Autonomous Vehicle Navigation, 2001
ISBN 3-7908-1341-9

Vol. 62. N. Baba and L.C. Jain (Eds.)
Computational Intelligence in Games, 2001
ISBN 3-7908-1348-6

Vol. 63. O. Castillo and P. Melin
Soft Computing for Control of Non-Linear Dynamical Systems, 2001
ISBN 3-7908-1349-4

Vol. 64. I. Nishizaki and M. Sakawa
Fuzzy and Multiobjective Games for Conflict Resolution, 2001
ISBN 3-7908-1341-9

Vol. 65. E. Orłowska and A. Szalas (Eds.)
Relational Methods for Computer Science Applications, 2001
ISBN 3-7908-1365-6

Vol. 66. R.J. Howlett and L.C. Jain (Eds.)
Radial Basis Function Networks 1, 2001
ISBN 3-7908-1367-2

Vol. 67. R.J. Howlett and L.C. Jain (Eds.)
Radial Basis Function Networks 2, 2001
ISBN 3-7908-1368-0

Vol. 68. A. Kandel, M. Last and H. Bunke (Eds.)
Data Mining and Computational Intelligence, 2001
ISBN 3-7908-1371-0

Vol. 69. A. Piegat
Fuzzy Modeling and Control, 2001
ISBN 3-7908-1385-0

Vol. 70. W. Pedrycz (Ed.)
Granular Computing, 2001
ISBN 3-7908-1387-7

Vol. 71. K. Leiviskä (Ed.)
Industrial Applications of Soft Computing, 2001
ISBN 3-7908-1388-5

Vol. 72. M. Mareš
Fuzzy Cooperative Games, 2001
ISBN 3-7908-1392-3

Vol. 73. Y. Yoshida (Ed.)
Dynamical Aspects in Fuzzy Decision, 2001
ISBN 3-7908-1397-4

Vol. 74. H.-N. Teodorescu, L.C. Jain and A. Kandel (Eds.)
Hardware Implementation of Intelligent Systems, 2001
ISBN 3-7908-1399-0

Vol. 75. V. Loia and S. Sessa (Eds.)
Soft Computing Agents, 2001
ISBN 3-7908-1404-0

Vol. 76. D. Ruan, J. Kacprzyk and M. Fedrizzi (Eds.)
Soft Computing for Risk Evaluation and Management, 2001
ISBN 3-7908-1406-7

Weiru Liu

Propositional, Probabilistic and Evidential Reasoning

Integrating Numerical and Symbolic Approaches

With 9 Figures
and 35 Tables

Springer-Verlag Berlin Heidelberg GmbH

Dr. Weiru Liu
School of Information and Software Engineering
University of Ulster at Jordanstown
Newtownabbey
Co. Antrim BT37 0QB
Northern Ireland
United Kingdom
w.liu@ulst.ac.uk

ISSN 1434-9922

DOI 10.1007/978-3-7908-1811-6

Cataloging-in-Publication Data applied for
Die Deutsche Bibliothek – CIP-Einheitsaufnahme
Liu, Weiru: Propositional, probabilistic and evidential reasoning: integrating numerical and symbolic approaches; with 35 tables / Weiru Liu. – Heidelberg; New York: Physica-Verl., 2001
(Studies in fuzziness and soft computing; Vol. 77)

Originally published by Physica-Verlag Heidelberg in 2001
MyCopy version of the original edition 2001

Hardcover Design: Erich Kirchner, Heidelberg
www.springer.com/mycopy

To my parents

Foreword

How to draw plausible conclusions from uncertain and conflicting sources of evidence is one of the major intellectual challenges of Artificial Intelligence. It is a prerequisite of the smart technology needed to help humans cope with the information explosion of the modern world. In addition, computational modelling of uncertain reasoning is a key to understanding human rationality.

Previous computational accounts of uncertain reasoning have fallen into two camps: purely symbolic and numeric. This book represents a major advance by presenting a unifying framework which unites these opposing camps. The Incidence Calculus can be viewed as both a symbolic and a numeric mechanism. Numeric values are assigned indirectly to evidence via the possible worlds in which that evidence is true. This facilitates purely symbolic reasoning using the possible worlds and numeric reasoning via the probabilities of those possible worlds. Moreover, the indirect assignment solves some difficult technical problems, like the combination of dependent sources of evidcence, which had defeated earlier mechanisms.

Weiru Liu generalises the Incidence Calculus and then compares it to a succession of earlier computational mechanisms for uncertain reasoning: Dempster-Shafer Theory, Assumption-Based Truth Maintenance, Probabilistic Logic, Rough Sets, etc. She shows how each of them is represented and interpreted in Incidence Calculus. The consequence is a unified mechanism which includes both symbolic and numeric mechanisms as special cases. It provides a bridge between symbolic and numeric approaches, retaining the advantages of both and overcoming some of their disadvantages.

This book promises to be an influential and significant milestone in uncertain reasoning and deserves to be widely read.

Edinburgh, UK Alan Bundy

Preface

The amazing and exciting power of common sense reasoning lies in the ability of human beings to draw rational conclusions from a collection of information, no matter how uncertain, imprecise, incomplete and vague the information is. This ability is certainly far beyond what classical logic can model. The demands for creating more sophisticated representation and reasoning paradigms have promoted the development of non-monotonic logics, default logic, model logics, truth maintenance systems, bayesian networks, fuzzy logic, possibility theory and belief function theory. All these different representation and reasoning mechanisms can be broadly divided into two main camps, purely symbolically based and numerically oriented.

It gradually became apparent that it is inadequate to model the reasoning procedure of human beings in either purely symbolic form or numerical structure alone. The extraordinary complex nature of common sense reasoning often requires the contribution of both symbolic and numerical reasoning elements. This understanding has invoked a string of attempts to unify these two reasoning mechanisms in order to achieve a more powerful combination. Probabilistic logic and probabilistic oriented assumption based truth maintenance system (ATMS) are just two examples of these various attempts.

This book aims at studying the nature of integrating numerical and symbolic approaches for reasoning under uncertainty. The contents of the book can be divided into two parts. The first part consists of a broad coverage and analysis of the attempts on the integration of the two streams of reasoning methods in the past decades, and materials on introduction to and modification of incidence calculus which is a mechanism possessing features of both numerical and symbolic reasoning systems. In the second part of the book, the Dempster-Shafer theory of evidence and the assumption-based truth maintenance system are selected as representatives of the two reasoning patterns. Both theories are critically reviewed before their inter-relationships with incidence calculus are fully examined. Finally, at the end of this part, rough sets theory and its natural connection with incidence calculus is presented, followed by a short summary concluding the book.

The major part of this work was carried out while the author was a PhD student in the Department of Artificial Intelligence, at the University of Edinburgh, sponsored by the Colin and Ethel Gordon Scholarship Award of the Faculty of Science and Engineering and an UK Overseas Research Studentship (ORS) Award, under the supervision of Alan Bundy and Dave Robertson, to whom I wish to express my sincere gratitude. A special thank goes to Alan Bundy and Dave McBryan for our collaborative work after I left Edinburgh. I would also like to thank following people (in alphabetical order) who have commented on different draft chapters during the preparation of the manuscript, David Bell, Alan Bundy, Dave Bustard, Ivo Düntsch, Jun Hong, Sally McClean, Michael McTear, Simon Parsons and Andrzej Skowron. In particular, Simon Parsons's suggestions on Chapter 1 and Ivo Düntsch and Andrzej Skowron's comments on Chapter 10 have helped improve these two chapters significantly. Finally, thanks should also go to the series editor, Janusz Kacprzyk, for his encouragement in publishing this book.

Some previous publications have contributed to the contents of various chapters. Sections 5.2 – 5.4 and 7.1 – 7.3 appeared in [Liu and Bundy, 1994], Sections 3.3 – 3.4 and Chapter 4 in [Liu *et al.*, 1998], Sections 6.2 and 6.3 in [Liu and Hong, 1999], and in Chapter 9 in [Liu and Bundy, 1996].

University of Ulster at Jordanstown
Northern Ireland

Weiru Liu

Contents

Chapter 1

Introduction

It has long been recognised that the problem of representing and reasoning with a large collection of knowledge and evidence (both of which can be called information) is a fundamental issue in Artificial Intelligence (AI), particularly, when this collection involves uncertainty. To make decisions based on this large amount of information, we need formal approaches that allow for rational reasoning to take place.

1.1 Classical and Non-monotonic Reasoning

In earlier artificial intelligence systems, a prime choice for representing and reasoning with certain information fell to *formal logic or classical logic*, such as propositional logic and first order logic. The use of logic provides a formal language with a well-defined syntax along with a syntactical relation describing formal consequences [Schaub, 1997].

It turns out that common sense reasoning by human beings cannot be fully addressed in classical logic. A vital part of common sense reasoning is to make use of available knowledge to derive conclusions which may subsequently be ruled out (or partially ruled out) when new information is available. This is known as *belief revision* [Gardenfors, 1992]. Therefore, a more adequate reasoning system is required to deal with *incomplete or uncertain* information. Various forms of *non-monotonic logic* have emerged to meet this need, such as McCarthy's *circumscription* [McCarthy, 1980], McDermott and Doyle's *non-monotonic logic I* [McDermott and Doyle, 1980], and Reiter's *default logic* [Reiter, 1980]. These systems are non-monotonic in the sense that newly arrived information may lead to a contradiction to formerly derived conclusions and the removal of invalid conclusions. This is a significant improvement over classical logic where the addition of any further information cannot rule out any previously derived conclusions.

Consider, for instance, the following scenario. Person A will have a meal with his colleagues tonight. His current knowledge about the location of the dinner is either *at restaurant P*, denoted as p or *at restaurant Q*, denoted as q, and he is sure that it has to be one of the two choices. Then his current knowledge base K includes statement $p \vee q$ *is true*, where $\vee$ is a disjunctive operation in logic. If A later learns that the dinner is in p, his knowledge base K will be modified by adding two more conclusions: p *is true* and q *is false*; otherwise, q *is true* and p *is false*. In either case, the conclusion $p \vee q$ *is true* is always correct. This reasoning procedure can be captured precisely in classical logic. However, if person A's knowledge about the dinner in either p or q does not exclude other possibilities, and subsequently A is informed that the dinner will be at *restaurant R*, denoted as r, then A's knowledge base K is revised so that $p \vee q$ *is false* and r *is true*. This reasoning procedure is beyond the capability of classical logic and can only be carried out in a non-monotonic reasoning system.

In fact, when person A does not hold absolute confidence in $p \vee q$ (e.g., he does not exclude any other possibility due to lack of information) or A is not certain that $p \vee q$ is definitely true (e.g., because A did not hear over the phone clearly), A's knowledge is either *incomplete* or *uncertain*. The incompleteness and uncertainty implied in a piece of information can be addressed using a numerical value (or a pair of values) which indicates the confidence we hold in this information. This is known as *reasoning under uncertainty*. Non-monotonic reasoning approaches mentioned above are able to cope with incomplete or uncertain information only when the incompleteness and uncertainty is not described in a numerical way.

To characterise the nature of both incompleteness and uncertainty in information, a numerical value is used to describe the degree of belief in it. The most traditional method is *probability theory*, such as attaching 0.9 to the statement $p \vee q$, meaning that the probability of having dinner at either P or Q is 0.9. The remaining 0.1 is assigned to the statement $\neg(p \vee q)$, where $\neg$ is the negation operation in logic,. However, one may argue that assigning the remaining 0.1 in this way does not bring out the real sense of information deficiency.

The Dempster-Shafer theory of evidence (DS theory) [Dempster, 1967], [Shafer, 1976] copes with this situation by assigning the remaining 0.1 to both $(p \vee q)$ *and* $\neg(p \vee q)$ meaning that it is not possible to assign 0.1 to any specific statement because the current information is not sufficient enough for an agent to decide how to do it. DS theory is often regarded as a more general numerical mechanism than probability theory because of its ability to represent incomplete information (or *ignorance*).

Certainly, the delicate relationship between DS theory and probability theory goes far beyond this issue (and also beyond the scope of this book). This was discussed in many papers as, e.g., [Lemmer, 1986], [Kyburg, 1987], [Black,1987], [Hunter,1987], [Halpern and Fagin, 1992] or [Voorbraak, 1991].

Other numerical calculi include *Bayesian belief networks* [Pearl, 1988], *fuzzy sets* [Zadeh, 1975], *possibility theory* [Dubois and Prade,1988], *the certainty factor model* [Shortliffe, 1976], etc. Numerical reasoning systems are special forms of non-monotonic reasoning, since an agent's confidence in a statement may change when more information is gathered.

1.2 Classifications of Non-monotonic Systems

A non-monotonic reasoning system (or even a monotonic system) comprises of two primary components: a facts/knowledge representation component and a reasoning component. If both the representation and reasoning components of a system are confined to a symbolic formalism, the systems is characterized as *purely symbolic* or simply *symbolic.* On the other hand, if both are in a numerical form, the system is called *purely numerical*, or simply *numerical*; otherwise, the system is *hybrid.* The first two categories are referred to as non-numerical and numerical methods in [Mamdani *et al.*, 1988] and [Bhatnagar and Kanal, 1986], respectively. As for hybrid systems, we further divide them into the four categories[1]:

- symbolic reasoning dominated systems where numerical measures are fitted on top of symbolic reasoning,
- numerical reasoning dominated systems where symbolic reasoning/conclusions are drawn on top of numerical measures,
- dual inference systems where uncertain calculi are fitted on top of symbolic reasoning, and inferences are carried out in both forms simultaneously,
- hybrid dual inference systems where the representation component, fitted on top of a symbolic reasoning component, consists of both numerical and qualitative measures in addition to the dual inference.

We shall look at each of these categories in detail later.

1.2.1 Purely Symbolic Non-monotonic Systems

The normal province of *purely symbolic non-monotonic systems*[2] is the derivation from initially precise information to a precise conclusion, although such a conclusion is understood to be tentative, it may have to be retracted after new information is added, such as default logic and the assumption-based truth maintenance system (the ATMS). In these reasoning systems, information (including both knowledge and evidence) and conclusions are all represented in a symbolic form. No numerical values are used to assess the truth value of any statement.

[1] I own this classification to Simon Parsons who made this suggestion to me.

[2] Simply *symbolic* in the rest of the book.

The logical soundness of symbolic approaches has been proved powerful in many aspects. The major drawback that these approaches suffer from is the inability to represent uncertain information. For instance, it is not possible to represent sentences like 'q is possibly true' or 'q is true with probability 0.7'.

1.2.2 Purely Numerical Non-monotonic Systems

Those mechanisms which represent and reason with uncertain or incomplete information using 'probabilities' or 'possibilities' belong to this category, for instance, probability theory and DS theory. Numerous *purely numerical*[3] approaches for managing uncertain information have been proposed as we mentioned earlier (and also see [Kruse *et al.*, 1992], [Krause and Clark, 1993] for details).

The common features of the reasoning mechanisms in this category lie in explicitly modelling or describing uncertain or incomplete information with numerical measurement and using the measurement to make further judgements. A reasoning procedure in this category involves both propagating and calculating numerical values of statements.

Although a method of this kind overcomes the drawback of symbolic non-monotonic reasoning systems, in terms of representing uncertain information, nevertheless, it raises its own problems, such as the difficulty of handling non-independent pieces of evidence when evidential combination is involved and the inability of explaining the real meaning of numerical values assigned to the hypotheses after a few steps of reasoning.

1.2.3 Hybrid Non-monotonic Systems

Both symbolic and numerical approaches in the first two categories have advantages and limitations. An intensive survey and discussion was provided in [Dubois and Prade, 1994a] between classical logic and the Bayesian networks. It concluded that

> "...the deficiencies of classical logic and of Bayesian networks with respect to the plausible reasoning endeavour are not the same. The overriding ambition for knowledge representation and reasoning in the domain of plausible inference is to identify a logic which combines the advantages of Bayesian networks with those of classical logic."

Taking Bayesian networks and classical logic as the typical representatives of numerical and symbolic reasoning techniques, Dubois and Prade's analysis can be generalised to pinpoint the very different nature of the two categories of reasoning patterns.

[3] Simply *numerical* in the rest of the book.

They also stated that "... numerical and symbolic approaches to uncertainty should not be considered as competing models. It is far more interesting and fruitful to display their underlying coherence."

Dubois and Prade's statement echos to an earlier summary made by Shafer and Pearl when they were introducing Chapter 9 – Integrating Probability and Logic, in their edited book 'Readings in Uncertain Reasoning' [Shafer and Pearl, 1990]:

> "... qualitative relationship can be seen as abstractions from probability ideas, and numerical probabilities can be seen as supplementary to traditionally symbolic systems."

Symbolic inference dominated hybrid systems: where uncertainty measures are fitted on top of symbolic reasoning. A system of this type keeps its reasoning component in a symbolic form. Uncertainty measures are used to calculate degrees of beliefs only at certain stages. A typical example is the integration of probability theory or DS theory into the ATMS, where the degree of belief of a statement (node) can be calculated only when the label of the node has been derived. The derivation of a label is purely symbolic.

Numerical inference dominated hybrid systems: where symbolic reasoning is fitted on top of numerical measures. We have not seen much work in this stream so far. One possible example to illustrate the nature of systems of this type could be Liu and Wellman's ([Liu and Wellman, 1998a]) work on solving ambiguous qualitative influence relationships[4] using numerical probabilistic distributions locally[5]. In their system, whenever an ambiguous qualitative relationship arises, a numerical method, either marginalizing nodes contributing to the ambiguity or estimating bounds of probability distributions of relevant nodes, is called in to achieve an unambiguous relationship.

Dual inference systems: where uncertainty calculi are fitted on top of symbolic reasoning and the inferences in both forms are tied together. These systems first integrate an uncertainty calculus into a symbolic reasoning mechanism, similar to systems of the fist type, then carry out inference simultaneously both in reasoning component symbolically and in representation component numerically or non-numerically. We further divide systems in this category into two groups, *symbolic oriented dual systems* or *numerical oriented dual systems*, where an uncertainty calculus in the former is non-numerical and in the latter is numerical.

A typical example of symbolic oriented dual systems would be Cohen's *endorsement* [Cohen, 1985]. In his system, linguistic words, such as *high*,

[4]See the introduction to qualitative probabilistic networks in Section 1.4.

[5]Strictly speaking, this example does not entirely fall into this category, since not all qualitative relationships are derived from numerical calculations. However, we cannot find a more suitable example belonging to this group.

and *low*, rather than numerical degrees or intervals are employed to represent vague and imprecise information or knowledge.

Examples of numerical oriented dual systems include probabilistic logic and possibilistic logic. In probabilistic logic, particularly in anytime deduction algorithms (e.g. [Frisch and Haddawy, 1994]), conventional inference rules in logic (such as if both $q_1 \rightarrow q_2$ and $q_2 \rightarrow q_3$ are true, then $q_1 \rightarrow q_3$ is true) are used to guide the inference path to infer new statements, which will then immediately lead to the calculation of probabilities (or intervals) on these newly derived statements. Subsequently, the probabilities (or intervals) on these new statements will be used to calculate probabilities on other statements in the next round and so on.

Possibilistic logic performs its dual inference in a similar fashion, except that the calculation of possibility and necessity measures are limited to mainly max and min operations (occassionally with $1 - max$ or $1 - min$), in contrast to real arithmetical calculations in probabilistic logic. We shall look at possibilistic logic in more details in a separate section later on.

Hybrid dual inference systems: where the uncertainty calculi consist of both numerical and non-numerical measures in dual inference systems. A hybrid dual system cannot be simply labelled as numerical oriented or symbolic oriented and it blurs the distinction of the two divisions. A good example of this type is an argumentation system in which some dictionaries use linguistic measures and others use numerical measures (see the introduction of argumentation later in the chapter). When all the dictionaries are defined in a numerical form, this argumentation system is numerical oriented dual system, while when all the dictionaries are in a non-numerical form, it is symbolic oriented.

1.3 Hybrid Systems: Some Examples

Symbolic and numerical approaches are, to some extent, complementary rather than exclusive. Therefore, attempts, as early as the beginning of the 80's (e.g. [?], [Ginsberg, 1984]), have been made to integrate them to solve complex problems. In [Rich, 1983] a likelihood-based interpretation of default rules was proposed using the certainty-factors calculus. DS theory was employed to describe default theory in [Ginsberg, 1984] and [Baldwin, 1987]. An extension of the classical logic (either propositional or first order) to a probabilistic one was given in [Bundy, 1985] and [Nilsson, 1986].

In [de Kleer and Williams, 1987] there were associated probabilities with assumptions in ATMS. The application of the method proposed by Carnap for the development of logical foundations of probability theory to *epistemic logic* structures was shown in [Ruspini, 1987].

In [Dubois and Prade 1987] possibilistic logic was developed by assigning possibilities and necessities in *possibility theory* to formulae in first-order logic [Dubois and Prade,1988a].

In [Provan, 1989], [d'Ambrosio, 1988] and [Laskey and Lehner, 1989] a formal relationship between DS theory and an ATMS was sought.

In [Dubois *et al.*, 1990] an extension of an ATMS in the framework of possibility and necessity measures was given. An attempt to extract numerical values from ordered beliefs was reported in [Cloteaux, et al, 1998].

These attempts have matured over the last 15 years (e.g., [Pearl, 1988], [Goldszmidt and Pearl, 1996], [Bacchus *et al.*, 1996], [Liu and Bundy, 1996], [Kohlas *et al.*, 1998], [Anrig and Monney, 1999]), in terms of theoretical studies of the nature of reasoning systems possessing both numerical and symbolic reasoning features.

1.3.1 Extending Classical Logic

The classical propositional logic, like any other symbolic reasoning mechanisms, lacks the ability to deal with uncertainty in information. Work has been done on the generalization of propositional logic or first order logic to probabilistic logic, that is, to extend a two-valued logic to a multiple-valued logic by assigning numerical values to sentences, to reflect an agent's belief.

Bundy's incidence calculus

Bundy [Bundy, 1985] introduced probabilities into propositional logic by assigning probabilities to sentences via a set of *possible worlds*. This approach makes the association of probabilities to sentences indirect, unlike Nilsson's method.

To calculate the probability of a sentence, one first has to obtain all the possible worlds which support the sentence. The advantage of this approach is that it is easy to calculate probabilities (or their bounds). For example, if we are told that a set of possible worlds, W_1, supports q_1, and another set of possible worlds, W_2, supports $q_1 \rightarrow q_2$, then it is possible to say that the lower bound of the support set for q_2 is $W_1 \cap W_2$. It is, then, easy to calculate the probability of $W_1 \cap W_2$.

The calculation of probabilities through a set of possible worlds is the main component of indirect encoding and is the key idea in incidence calculus. The formal connections of incidence calculus with interval structures and Dempster-Shafer theory of evidence were reported in [Wong and Wang, 1993] and [Correa da Silva and Bundy, 1990].

Nilsson's probabilistic logic

Nilsson [Nilsson, 1986] combined propositional logic with probability theory through assigning probabilities to sentences directly (called *direct encoding*). That is, the truth value of a sentence is a probability value rather than just truth or false; for example $prob(p) = 0.7$ where p is a sentence. The source of a probability assigned to a sentence is explained as follows. Given a sentence, say ϕ, if we start with a set of samples (as required in probability theory), this sentence can be either true or false on one sample. These samples are called possible worlds. So there are two sets of possible worlds, ϕ is true in one of them and false in another. Therefore, Nilsson concluded that "the probability of a sentence is the sum of the probabilities of the sets of possible worlds in which that sentence is true".

Apart from its obvious merit of representing uncertainty, this approach suffers from the difficulty of propagating probabilities based on an initial probability assignment. For instance, if we know that $prob(q_1) = 0.7$ and $prob(q_1 \rightarrow q_2) = 0.5$ where $\rightarrow$ is an implication operation in logic, it is not possible to calculate the probability of q_2; we can only know that the probability of q_2 lies in the interval

$$prob(q_1) + prob(q_1 \rightarrow q_2) - 1, prob(q_1 \rightarrow q_2)].$$

For more complicated cases, it is even difficult to tell the lower or upper bound of probabilities on sentences. This situation was picked up by Snow [Snow, 1991] who suggested the use of a three valued logic (the truth value of a sentence in a particular possible world can be 'true', 'false', or 'unknown') to construct a 'compressed' constrained system for generating entailments.

Obviously, Nilsson's approach fails when $prob(q_1) + prob(q_1 \rightarrow q_2) < 1$ occurs. McLeish [McLeish, 1989] made two extensions of Nilsson's method:

- by allowing $prob(q_1) + prob(q_1 \rightarrow q_2) < 1$ to happen (this extension is especially useful for the representation of default information with conflict),
- by replacing probability assignments on statements with belief functions in DS theory.

In a separate article by McLeish [McLeish, 1988], the discussion focused on how inconsistent information may be dealt with in probabilistic logic and how probabilistic entailment may be carried out.

In Grosof's paper [Grosof, 1986] Nilsson's work was extended with upper and lower bounds on conditional probabilities, as the author believed that the bounds were especially useful for describing the semantics of probabilistic knowledge, for describing intermediate states of probabilistic inference and updating. This inequality paradigm, a generalized probabilistic logic, was also claimed to take DS theory as a special case.

A method for probabilistic reasoning extending Nilsson's work by incorporating conditional probabilities was proposed by Frisch and Haddawy [Frisch and Haddawy, 1994]. In this approach, a set of inference rules is created to calculate probabilistic entailments in the form of intervals. An *anytime deduction algorithm* is designed to compute an increasingly tightened interval until the interval concerning the probability range of any entailed sentence is fine enough for the problem being considered. A distinctive feature of this method is that the algorithm can stop at any time to yield a solution (may be a loose interval) if this solution is good enough.

Bacchus's work

Bacchus [Bacchus, 1988], [Bacchus, 1990] examined Nilsson's probabilistic logic and other extensions (e.g., [Carnap, 1962], [Scott and Krauss, 1966], [Field, 1977], [LeBlanc, 1983], and [Fagin *et al.*, 1988]) with a different perspective. He particularly pointed out the limitation of the method of assigning probabilities to logical sentences through the probability distribution over a set of possible worlds in respect of statistical assertions. This limitation leads to the difficulty of explaining the semantics of probabilities assigned to sentences involved in a default reasoning system and the difficulty of representing statistical knowledge such as 'more than 90% of all birds fly'.

Another aspect that Bacchus addressed is the powerful expressiveness that first order probabilistic logic has over propositional probabilistic logic. He then formulated a solution in the form of first order logic with its adaptation to a situation where statistical knowledge and defaults inference are used.

For example, in the language **Lp** he proposed, an agent's knowledge 'John probably has some type of cancer' is expressed as

$$prob(\exists x.has\_cancer\_type(John, x)) > 0.5.$$

On the other hand, a piece of statistical information 'more than 50% of all dogs bark' has to be written as

$$[Bark(x) \mid Dog(x)]_x > 0.5$$

which means that there is a chance of over 50% that a randomly selected dog x can bark.

In his language, square brackets are used to bind free variables to open formulae to form *probability terms*, such as, $[Bark(x) \mid Dog(x)]_x$.

Bacchus's opinion about differentiating statistical knowledge from other subjective/objective knowledge has been further studied in a number of other papers (e.g., [Halpern, 1990], [Bacchus *et al.*, 1994], [Bacchus *et al.*, 1996]).

In [Bacchus *et al.*, 1996], a random-worlds method is used to integrate qualitative default reasoning with quantitative probabilistic reasoning by providing a language in which both types of information can be easily expressed. The language in this method is a variant of **Lp** in which *approximate equality* replaces equality when expressing some statistical or default knowledge. For instance, an assertion that 'about 80% of patients with jaundice have hepatitis' will be expressed as $[Hep(x) \mid Jaun(x)]_x \approx 0.9$ instead of $[Hep(x) \mid Jaun(x)]_x = 0.8$, which says that the proportion of jaundiced patients with hepatitis is close to 80% within some tolerance τ of 0.8.

Other work on extending classical logic

An attempt to combine formal logics with uncertainty calculi, other than probability theory, has been reported in [Saffiotti, 1992] in which a belief function logic was formalised. In this framework a pair of values $[b, p]$ may be assigned to a first order logic sentence where b is the degree of belief that the sentence is absolutely true and p is the degree of belief that the sentence is possibly true. A corresponding technique of handling partial inconsistency and contradiction is formalised as well.

A probabilistic extension of terminological logics was presented in the works [Heinsohn, 1991] and [Heinsohn, 1994]. The language *ALLP* can handle a variety of forms of knowledge including terminological knowledgecovering term descriptions and uncertain knowledge.

1.3.2 Associating Numerical Values with Non-monotonic Rules

The research results we are going to discuss in this subsection can be, in principle, broadly classified as extending classical logic and therefore can be included in the above subsection. However, we prefer to separate them out because these research results address a set of specific logical rules where exceptions need to be considered.

Associating non-monotonic rules with certainty factors

Rich [Rich, 1983] introduced certainty factors (CFs) into *semantic nets* [Quillian, 1968] to represent the degree of belief held in properties they represent. As a result, the well-known non-monotonic rule *birds fly, ostriches do not fly* widely used in the literature is represented as

$$birds \rightarrow fly : (CF = 0.95). \qquad ostriches \rightarrow fly : (CF = -1.0). \tag{1.1}$$

The certainty factors 0.95 and -1.0 indicate that 95% birds fly and ostriches do not fly at all, respectively.

A CF in (-1, 1) indicates that either the evidence supports or refutes a proposition, with some degree of uncertainty, if it is not known definitively true or false. Otherwise the CF would be either 1 or -1 for absolutely true or false.

Associating non-monotonic rules with DS theory

Ginsberg [Ginsberg, 1984] argued that using a range rather than an exact value may be more reasonable to reflect the real meaning implied in *birds fly* – it is better to believe that between 90% and 98% of birds fly instead of exactly 90% do. Thus, equation (1.1) becomes

$$birds \rightarrow fly : (0.9, 0.02), \qquad ostriches \rightarrow fly : (0.0, 1.0). \tag{1.2}$$

The pair (a, b) with a default rule indicates that to the degree a we believe this rule and to b we disbelieve it. Therefore, the maximum extent to which we believe the rule is $1 - b$. *Dempster's combination rule* in DS theory [Shafer, 1976] is used to combine multiple non-monotonic rules with the same conclusions.

McLeish [McLeish, 1990] further investigated the application of Dempster's combination rule in a non-monotonic reasoning system, with particular attention paid to the situation when conflict non-monotonic rules are involved. McLeish argued that the erroneous results of applying Dempster's combination rule to a non-monotonic situation, presented by Pearl in [Pearl, 1988] using $\epsilon-$calculus formulation, could be avoided if the frames of discernment involved are chosen more carefully.

An approach to dealing with default information using DS theory was also investigated in [Benferhat *et al.*, 1995] in which ϵ-belief functions are used to represent non-monotonic information and the least-commitment principle is used to define non-monotonic consequence relations.

1.3.3 Extending the ATMS

Assumption based truth maintenance systems (ATMSs) [de Kleer, 1986a] were stimulated by Doyle's work on truth maintenance systems (TMSs) [Doyle, 1979]. In an ATMS, dependent relations on statements (nodes or sentences) are explicitly recorded and maintained.

Assumptions are special statements which are assumed to be true without requiring any extra information, when there are no conflicts. The truth of any other statement (except facts, called *premises* in an ATMS) is supported by sets of assumptions, the collection of which is known as the *label* of the statement. Justifications, provided by the system designer, are the sources of logical reasons to derive labels.

Table 1.1 reveals how labels, containing only assumptions (all in capital letters) are derived through justifications.

$q_1 \rightarrow q_2$ and $q_2 \rightarrow q_3$ are called *assumed nodes* with justifications mentioning only assumptions, meaning they hold under A and B respectively. q_1 is called a premise which is observed to be true. N_4 to N_6 are *derived nodes* with labels determined from given justifications.

In the traditional ATMS, assumptions can only be *true* or *false*. The integration of numerical approaches into an ATMS means that a numerical value is attached to an assumption to indicate to what degree the agent believes in this assumption, so as to calculate the beliefs on sentences as a consequence.

Table 1.1: An ATMS example

Node	Statement	Label	Justification
N_1	$q_1 \rightarrow q_2$	$\{\{A\}\}$	$\{(A)\}$
N_2	$q_2 \rightarrow q_3$	$\{\{B\}\}$	$\{(B)\}$
N_3	q_1	$\{\}$	$\{\}$
N_4	q_2	$\{\{A\}\}$	$\{(N_1, N_3)\}$
N_5	q_3	$\{\{A, B\}\}$	$\{(N_1, N_2, N_3)\}$
N_6	$q_1 \rightarrow q_3$	$\{\{A, B\}\}$	$\{(N_1, N_2)\}$

Extending the ATMS via DS theory

Possible connections/combinations between the ATMS and Dempster-Shafer theory of evidence are studied in many works (e.g., [Provan, 1989], [d'Ambrosio, 1988], [Laskey and Lehner, 1989], etc.).

In [d'Ambrosio, 1988], a hybrid reasoning scheme that combines symbolic and numerical methods for uncertainty management is provided. The hybrid is based on symbolic techniques adapted from the ATMS, combined with numerical methods adapted from DS theory [d'Ambrosio, 1988]. The most important advantages of such a combination are its improved management of dependent and partially independent evidence and its ability to query an agent's belief in a proposition from multiple sources.

In [Laskey and Lehner, 1989] a the formal connection of these two approaches was investigated to a further extent:

> "A formal equivalence is demonstrated between Shafer-Dempster belief theory and assumption-based truth maintenance with a probability calculus on the assumptions. This equivalence means that any Shafer-Dempster inference network can be represented as a set of ATMS justifications with probabilities attached to assumptions. ... The approach described here unifies symbolic and numerical approaches to uncertainty management."

Extending ATMS via probability theory

De Kleer and Williams [de Kleer and Williams, 1987] were the first among various efforts in associating probabilities with assumptions in an ATMS. They developed a probabilistic ATMS for fault management in electronic circuit design.

The most recent work on probabilistic ATMS [Kohlas *et al.*, 1998] for dealing with problems in diagnosis can be explained as a further effort to integrate ATMS with a numerical approach, although Reiter's idea on model-based diagnosis also plays a key role in it.

Based on the ideas of probabilistic assumption based reasoning and ATMS ([Provan, 1989], [Laskey and Lehner, 1989] and [Kohlas and Monney, 1993]),

Kohlas et al. developed a theory of diagnosis where each component is assigned to a prior probability indicating the probability that the component is working, before any observation is made. For every system state $x = \{x_1, \ldots, x_n\}$ of a system with n components (where x_j has a value out of domain $\{0, 1\}$ whose values stand for component j not working and working respectively), its prior probability is calculated. Once some observations are made, the diagnostic system identifies a set of system states which are compatible with the observations. This collection of possible explanations of the observations, N_d, is represented as a logical formula in the propositional language. The probability of N_d and the posterior probability of each state in N_d will then be calculated.

In summary, the main feature of the theory is that it is possible to determine efficiently arguments (system status) in favour of a diagnosis when a fault(s) is detected. Although these arguments represent the symbolic aspect of the diagnosis process themselves, on the numerical aspect, they can also be used to calculate posterior probabilities.

Extending ATMS via other numerical calculi

Dubois et al. [Dubois *et al.*, 1990] extended an ATMS to handle uncertain information in the context of possibility and necessity measures.

Fringuelli et al. [Fringuelli *et al.*, 1991] described a fuzzy truth maintenance system, an extension of an ATMS through fuzzy logic.

1.3.4 Possibilistic Logic

There has been a considerable research on integrating possibility theory [Dubois and Prade,1988a] with classical logic. Dubois and Prade first extended first-order logic by attaching either a possibility measure or a necessity measure to every first order logical formula to handle uncertainty [Dubois and Prade 1987], [Dubois and Prade, 1994b].

Possibility theory: In order to examine possibilistic logic, we start with basic notions in possibility theory, a theory developed based on the notion of fuzzy sets [Zadeh, 1975].

There are two fundamental set functions in possibility theory:

- *possibility measures* Π, and
- *necessity measures* N.

Given a set of events S, which contains the true event an agent is interested in and a subset A of S ($A \subseteq S$), the agent's confidence of believing that the true event is also in A can be expressed by a pair of values $\Pi(A)$ and $N(A)$, where $\Pi(A), N(A) \in [0, 1]$. $\Pi(A)$ estimates to what extent A contains the true event, or in other words, A is true, and $N(A)$ evaluates the degree of necessity that A is true.

Therefore, given any two subset A and B of S, the following equations hold.

$$\begin{aligned}\Pi(S) &= 1,\\ \Pi(\emptyset) &= 0,\\ \Pi(A \cup B) &= max(\Pi(A), \Pi(B)),\\ N(A \cap B) &= min(N(A), N(B)),\\ max(\Pi(A), \Pi(\bar{A})) &= 1,\\ min(N(A), N(\bar{A})) &= 0,\\ N(A) &= 1 - \Pi(\bar{A}).\end{aligned}$$

where $\bar{A}$ is the complementary set of A.

The first pair of equations suggests that the reference set S is always believed to contain the true event and the true event is never in the empty set.

The second pair states that for any two arbitrarily selected subsets, the possibility of the union set is true is not greater than that of the more possible subset among the two; on the other hand, the degree of necessity of their common subset is true is not smaller than that of the less possible one.

The third pair reveals that an agent can believe that at least one of the two contradictory events is completely possible; plus, when one event is possible, it does not rule out that its contrary being possible, however, the two contradictory events cannot share the slightest necessary (condition) simultaneously.

Finally, the last equation says "an event is necessary when its contrary is impossible" [Dubois and Prade,1988a].

Besides, for any $A \subseteq S$, $\Pi(A) \geq N(A)$, which indicates that an agent's perception of believing in an event is always greater (or at least equals to) than the actual degree of necessity of that event being true.

Possibilistic logic: Possibilistic logic, no matter whether built on propositional logic or first-order logic, is a logic capable of representing partial ignorance. In this book, we only consider propositional possibilistic logic.

Let P be a set of atomic propositions, each proposition q_j in P is assigned to a degree of possibility $\Pi(q_j)$ and a degree of necessity $N(q_j) = 1 - \Pi(\neg q_j)$ to measure the statement 'q_j is true'. For $q_j \in P$, if $N(q_j) = 1$ then q_j is true under all possible interpretations, if $\Pi(q_j) = 0$ then q_j is impossible (false) under all interpretations.

Therefore, $N(true) = 1$ and $\Pi(false) = 0$ echo to the convention in logic that sentence *true* is always true and sentence *false* is always false.

For any two atomic propositions q_j and q_l,

$$\Pi(q_j \vee q_l) = max(\Pi(q_j), \Pi(q_l)). \tag{1.3}$$

So, $N(\neg(q_j \vee q_l)) = 1 - \Pi(q_j \vee q_l) = 1 - max(\Pi(q_j), \Pi(q_l))$. On the other hand, $N(\neg(q_j \vee q_l)) = min(N(\neg q_j), N(\neg q_l))$. Therefore, if

$$1 - max(\Pi(q_j), \Pi(q_l)) \neq min(N(\neg q_j), N(\neg q_l)),$$

then the initial possibility-necessity assignment on P is not consistent.

In addition, the basic inference patterns, *modus ponens*, *modus tollens*, and *resolution principle* have been revised to [Dubois and Prade,1988a]:

$$\begin{aligned}
\Pi(q_l) \geq N(q_l) &\geq min(N(q_j), N(q_j \rightarrow q_l)), \\
N(q_j) \leq \Pi(q_j) &\leq max(\Pi(q_l), 1 - N(q_j \rightarrow q_l)), \\
N(q_l \vee q_r) &\geq min(N(q_j \vee q_l), N(\neg q_j \vee q_r)).
\end{aligned}$$

Dual nature of possibilistic logic: As we have briefly discussed in Section 1.2, inferences in possibilistic logic are carried out simultaneously both in the symbolic form (logical entailments) and in the numerical form (lower or upper bounds of necessity and possibility measures). For instance, if both sentences q_j and $q_j \rightarrow q_l$ are known to be true, then p_l is also true based on the *modus ponens* law. In possibilistic logic, this logical entailment can be used to guide the estimation of the lower bound of the necessity measure of q_l given $N(q_j)$ and $N(q_j \rightarrow q_l)$.

It is also worth to point out again that although both probabilistic logic and possibilistic logic make use of logical inference rules to guide their process of calculating numerical values (either probabilities, possibilities, necessities, or their bounds) on sentences, the actual calculation in the former usually requires addition and multiplication operations while the actual calculation in the latter performs mainly comparison operations among the numbers. It is because of this difference, possibilistic logic is also characterised as a "*quasi-qualitative* calculus" [Dubois and Prade,1988]. Numerical values are used to represent levels (grades) of beliefs in statements, they don't have to be precise.

Other topics such as theorem proving, automated reasoning, and resolution principles under uncertainty in possibilistic logic were also investigated in [Dubois *et al.*, 1989], [Dubois *et al.*, 1987], and [Dubois and Prade 1990b], respectively.

In [Boldrin and Sossai, 1995], a dynamic possibilistic logic was defined where two new connectives ($\neg$ for negation) and ($\otimes$ for a new type of conjunction) were introduced. As a consequence, the new conjunction offers a dynamic combination mechanism for combining information from different sources.

In a recent paper by Benferhat and Sossai [Benferhat and Sossai, 1998], the problem of merging several knowledge bases from different sources in the possibilistic logic framework was addressed. A set of combination rules were presented in the extended version of possibilistic logic.

1.3.5 Argumentation

Argumentation is a process of constructing *arguments* about possible consequences (or conclusions or claims) and assigning confidence to those consequences, based on their supporting arguments ([Fox, et al, 1993]). In a standard logic L, such as, propositional logic, an argument, denoted as, $q_1, \ldots, q_n \vdash_L q$, represents a sequence of inferences on $q_1, \ldots, q_n$ leading to a conclusion, using inferences rules in L. A correct argument leads to a valid conclusion.

As an argument involves a sequence of inferences employing different inference rules to derive a conclusion, it can be further expressed as (G, R, θ) for conclusion θ, based on supporting arguments in G, using rules in R [McBurney and Parsons, 2000].

Formally, $G = (\Theta_0, \theta_1, \Theta_1, \theta_2, \ldots, \theta_{n-1}, \Theta_{n-1})$ is an order sequence of well formed formulae (wffs) θ_j and possibly empty sets of wffs Θ_j $(n \geq 1)$ and $R = (\vdash_1, \vdash_2, \ldots, \vdash_n)$ an ordered sequence of inference rules such that:

$$\begin{aligned} \Theta_0 &\vdash_1 \theta_1, \\ \theta_1, \Theta_1 &\vdash_2 \theta_2, \\ &\vdots \\ \theta_{n-1}, \Theta_{n-1} &\vdash_n \theta. \end{aligned}$$

For instance, when $G = (\{q_1, q_1 \to q_2\}, q_2, \{q_2 \to q_3\})$ and $R = (a \wedge (a \to b) \models b, a \wedge (a \to b) \models b)$, then (G, R, q_3) is an argument for claim q_3.

In particular, the set $\{\theta_{k-1}\} \cup \Theta_{k-1}$ is called the *grounds* for θ_k and the conclusion θ is named the *claim* of this grounded argument. Since argumentation is designated to the representation of uncertainty in information rather than to express logical consequences in a different form, degrees of confidence in grounded arguments and in claims can be explicitly represented using values in a pre-defined set D.

Set D is known as a *dictionary* and its elements can either be quantitative (e.g., probabilities, possibilities) or qualitative (e.g. possible, highly possible, impossible). Moreover, a dictionary D_C for claims may be different from a dictionary D_G for grounds. In addition, a dictionary D_I for inference rules is also possible. For example, a D_I can have elements *applicable, sometimes applicable, not applicable.* Appropriate combination mechanisms about the degrees of confidence on grounds and on claims, as well as on inference rules should be designed in order to compute the effect of an argument on a claim or a set of arguments on the same claim.

When all the dictionaries in an argumentation system consist of qualitative measurement elements, then this system should be classified as *symbolic oriented dual system,* like the endorsement which we treat as a special form of argumentation. When all the dictionaries contain numerical values in an argumentation system, it should be classified as a *numerical oriented dual system.* Because dictionaries in an argumentation system are not necessarily confined to a single form (either qualitative or numerical), one dictionary may be in a numerical form and another in a qualitative form. A system of this kind is a hybrid dual system.

1.4 Qualitative Probabilistic Reasoning

Although, strictly speaking, the *qualitative probabilistic network* (QPN) (cf. [Wellman, 1990a]) is not a product of merging the classical logic with numerical calculi, we still prefer to mention this stream of work here briefly. This is because some of the work on qualitative representation of probabilistic knowledge in intelligent systems makes use of numerical values in the handling and deriving of qualitative relationships (e.g., [Liu and Wellman, 1998a]), and some works on symbolic probabilistic inference in Bayesian networks (e.g., [Shachter *et al.*, 1990], [Chang and Fung, 1991a]) represent probabilities (or conditional probabilities) symbolically . In the following, we shall distinguish two types of QPNs, *purely qualitative* and *semi-qualitative.*

1.4.1 Purely Qualitative Probabilistic Networks

Bayesian probabilistic networks [Pearl, 1988] require full numerical conditional probabilities being specified among nodes. This requirement is impractical in many applications where either conditional probabilities is unavailable or not necessary.

Purely qualitative probabilistic networks are qualitative abstractions of Bayesian probabilistic networks, in which the conditional probabilities are replaced by qualitative notations of influences. Therefore, a purely qualitative probabilistic network also consists of a set of nodes and a set of arcs among pairs of nodes. Each arc (with a direction) represents a possible influence relationship between two connected nodes, with '+' for a positive influence, '-' for a negative influence, '0' for no influence, and '?' for an undecidable (ambiguous) influence. For example, *smoking increases lung cancer* and *sun-cream will reduce the chance of skin cancer* can be respectively represented as an arc with '+' from node 'smoking' to node 'lung cancer', and an arc with '-' from node 'sun-cream' to node 'skin cancer'. Usually, a QPN does not have arcs with qualitative sign '0', since such a sign between a pair of connected nodes implies conditional independence among these two node, and no direct link will be needed.

Upon a well established QPN, every arc is labelled with one of the three signs, '+', '-', and '?', and every node with '0', initially. When a piece of evidence is received, the signs on corresponding nodes will be re-defined (either to be '+' or '-') and these revised signs will be propagated along the network using:

- the *symmetry property,*

Table 1.2: Qualitative sign multiplication

$\otimes$	+	−	0	?
+	+	−	0	?
−	−	+	0	?
0	0	0	0	0
?	?	?	0	?

Table 1.3: Qualitative sign addition

$\oplus$	+	−	0	?
+	+	?	+	?
−	?	−	−	?
0	+	−	0	?
?	?	?	?	?

- the *transitivity property*, and
- the *composition property.*

The *symmetry property* guarantees that if a direct arc from node A to node B bears an influence sign 'l' (l is one of the four possible signs), then sign 'l' from node B to node A can also be used in propagation, although, the arc from B to A is not explicitly represented in the network.

The *transitivity property* allows multiple signs along a single path between two nodes to be combined into a single influence bearing impact of these signs.

The *composition property* asserts that multiple influences between two nodes along different chains can be combined to a single compound influence. This property can also combine multiple influences on a single nodes from different sources.

Rules in the multiplication table (Table 1.2, [Wellman, 1990b]) tell exactly what compound influence, on the second node of a pair, shall be achieved after multiple signs on a single chain are combined, when the sign on the first node of the pair is known.

Rules in the addition table (Table 1.3, [Wellman, 1990b]) combine the signs derived on the same node from several arcs pointing to that node, to achieve a final influence from different sources.

In [Druzdzel and Henrion, 1993] an elegant sign propagation algorithm was developed. The main idea of their algorithm is to trace the effect of the values of observed nodes on other nodes in a network based on the properties of symmetry, transitivity and composition, as well as sign changing principles outlined in Tables 1.2 and 1.3.

One task with the qualitative probabilistic reasoning is to solve an ambiguous qualitative influence (known as a *trade-off*) either achieved along a single trail or achieved after combining several signs from different sources. Refining or enhancing an existing QPN to minimize the occurrence of '?' is one of the possible solutions to this problem.

In [Parsons, 1995], a refinement method is proposed in order to efficiently predict which node(s) is most likely to change given certain evidence and how a node's probability (sign) is going to change given two almost conflicting pieces of evidence. Three possible solutions were examined: (1) *the identification of extreme probabilities*, (2) *relative order of magnitude reasoning*, and (3) *absolute order of magnitude scheme*. In the identification of extreme probabilities (or *the categorical influences method*), a categorical influence of node A on B will either increase the probability of the highest value of B to 1 (denoted by sign '++') or decrease the probability of the highest value of B to 0 (denoted by sign '−−'), given that the probability of A with highest value is close to 1. Certainly, this method only solves some extreme cases.

The relative order of magnitude approach solves problems of conflicting influences but is less impressive for conflicting hypotheses. The absolute order of magnitude scheme address both conflicting influences and hypotheses satisfactorily.

The idea of identifying the strength of an influence, e.g., the categorical influences, was further developed in [Renooij and van der Gaag, 1999], where strong and week influences were clearly divided. An influence is named a strong influence (either positive '++' or negative '−−') if the degree of influence of one node on another is above a certain level (called a cut-off value). In this enhanced QPN, $\otimes$ and $\oplus$ operations were extended to allow strong and week influences to be combined either along a single trail or along different trails. However, the symmetry property does not hold anymore in this network. So an arc with a strong influence sign, say '++', will have an ambiguous positive influence (sign '+?') on the reverse arc. Same principle applies to strong negative influences.

Instead of refining a QPN in order to solve ambiguous influences, a different approach to resolve conflicts was adopted in [Renooij, et al, 2000]. This approach is built upon an existing QPN, with attention paid to the selection of nodes which have signs '?' after certain steps of propagation and which will pass this sign on to the node of interest. This type of nodes are named *pivotal nodes*. Pivotal nodes, other than the nodes of interest themselves, exist only when the fragment of the underlying graph satisfy certain conditions. When the condition is not met, their proposed algorithm cannot be applied.

1.4.2 Semi-qualitative Probabilistic Networks

Unlike the methods mentioned in the previous subsection, the approaches proposed in [Liu and Wellman, 1998a] make use of numerical probabilities provided by the network to help solving ambiguous influences in qualitative reasoning.

The two approaches:

- *the marginalization approach*, and
- *the state-space abstraction approach*,

both apply numerical reasoning to either subproblems or simplified versions of the original network to some point, in order to produce less ambiguous relationships in intermediate or final models.

The former incrementally marginalizes nodes that contribute to the ambiguous relationshps and the latter evaluates approximate Bayesian networks for bounds of a probability distribution and then use these bounds to determine qualitative influences in question.

In order to use either of the two approaches, a QPN should also possess sufficient numerical conditional probability distributions. Therefore, a QPN of this kind is no-longer a pure qualitative one, we say it is *semi-qualitative.* Therefore, such a QPN is a semi-qualitative probabilistic network (semi-QPN).

Apart from the qualitative probabilistic reasoning approach introduced above, an alternative approach to qualitative probabilities is the kappa-calculus (or *κ-calculus*)[6] developed in [Goldszmidt and Pearl, 1992] and originated from [Spohn, 1990].

In kappa-calculus, all probabilities (either conditional or prior/posterior) in a Bayesian belief network are represented by ϵ^k. k is an integral power of ϵ called a k ranking and ϵ (a threshold probability) is a variable whose value falls in [0,1] and is chosen dynamically according to reasoning results.

The relationship between a precise probability value p and a κ-ranking κ, given a particular ϵ, is

$$\epsilon^k > p \geq \epsilon^{k+1}. \tag{1.4}$$

For example, if the probability of a proposition (or event) q_1 is 0.001, then the κ ranking of q_1 is 2, when ϵ is chosen as 0.1. However, when the probability of proposition q_1 is changed to 0.1, its κ ranking is reduced to 1, under the condition that ϵ still has value 0.1. Using equation (1.4), it is possible to convert any probability value into a κ ranking and vice versa.

Propositions with smaller κ values are more possible compared to propositions with larger κ values. Given two propositions A and B and their κ rankings, the κ rankings of compound propositions $A \vee B$ and $A \mid B$ are calculated by:

$$\kappa(A \vee B) = min(\kappa(A), \kappa(B)),$$

and

$$\kappa(A \mid B) = \kappa(A \wedge B) - \kappa(B).$$

[6] We put kappa-calculus in this subsection, because its values of ϵ and rankings are numerical.

As κ rankings can be abstracted from probabilities and then be propagated in a causal network, it would be interesting to see whether the inference results of the original full probabilities and the abstracted κ rankings would agree with each other.

Performance analysis of applying kappa-calculus and numerical probabilities to practical problems was carried out in [Henrion *et al.*, 1994] and [Darwiche and Goldszmidt, 1994]. In [Henrion *et al.*, 1994] these two approaches were both applied to the *car would not start* scenario. The result shows that kappa-calculus may be reasonably reliable for diagnosis tasks with very small prior fault probabilities. However, for large average fault probabilities, the relative performance drops sharply. Certainly this is just one sample example, more rigorous and robust experimental results are needed in order to have a more accurate comparison between these two approaches.

Also, in [Darwiche and Goldszmidt, 1994], issues of abstracting kappa rankings from probabilities and the performance of abstracted κ causal networks were both examined. Three sets of experimental tests (ϵ was given values 0.2, 0.02, and 0.002 respectively) were conducted to illustrate the closeness of the final inference results obtained in an abstracted kappa model and in the original Bayesian belief network, although occasionally kappa calculus may disagree with probabilistic inference.

1.4.3 Incorporating Qualitative Reasoning into Quantitative Reasoning

Qualitative probabilistic reasoning has been used to explain inference results in a Bayesian network, as discussed in [Henrion and Druzdzel, 1990] and [Wellman, 1993]. It has also been employed to bound probability distributions as shown in [Liu and Wellman, 1998b].

In the works by [Shachter *et al.*, 1990], [Chang and Fung, 1991a], and [Chang and Fung, 1991b] it was shown how to perform symbolic probabilistic inference to answer general queries based on a given Bayesian network. Unlike the conventional probability propagation procedure used in traditional Bayesian networks where the major task is to calculate posterior probabilities of each node, symbolic probabilistic inference addresses the usage of symbolic notations of probabilities (or conditional probabilities) (e.g., $p(x)$, $p(x) \mid p(y)$) rather than the actual numerical values in the inference process. The actual calculation of posterior probabilities takes place only when an agent believes that a query has been sufficiently expressed in the symbolic form.

For example, in [Chang and Fung, 1991b] if an agent wants to know $p(XS \mid LE)$ to answer a query, based on a given Bayesian network. If, according to the network, $p(XS \mid LE)$ can be calculated through $\Sigma_B p(S \mid BL)p(X \mid EB)p(B \mid LE)$ where $p(X \mid EB)$ can be further calculated from $\Sigma_D p(X \mid E)p(D \mid BE)$, then $p(XS \mid LE)$ will be expressed as

$$\Sigma_B p(S \mid BL)(\Sigma_D p(X \mid E)p(D \mid BE))p(B \mid LE).$$

It is the symbolic form of $p(X \mid EB)$ rather than its actual value that will be used in the rest of the inference.

1.5 Rough Sets Theory

Rough sets theory ([Pawlak, 1982], [Komorowski *et al.*, 1999]) is a symbolic calculus which classifies objects of a non-empty *universe* into partitions via equivalence relations.

An equivalence relation divides objects of a universe into a list of disjoint subsets, with elements in the same subset indiscernible. When a category (also called a *concept*) cannot be precisely described by a subset of a set of objects (a universe), it is characterized with a pair of lower and upper approximations. The lower approximation is a subset containing those objects definitely supporting the category, while the upper approximation is a subset consisting of those objects which may support the category.

For example, assume that Table 1.4 contains the objects in U, and R is an equivalence relation on U which classifies objects according to their weight. Then $U/R = \{\{u_1, u_3, u_6\}, \{u_2, u_4, u_5\}\}$ defines two equivalence classes, each of which contains objects having the same value of 'weight'. Equivalence relation R can alternatively be defined as

$$u_j R u_l \Longleftrightarrow weight(u_j) = weight(u_l),$$

where $weight(x)$ will return the value of attribute 'weight' of object x.

Therefore, $U/R_{\{shape,size\}} = \{\{u_1, u_3, u_6\}, \{u_2\}, \{u_4\}, \{u_5\}\}$ will classify the objects according to their shapes and sizes. Similarly $U/R_{colour} = \{\{u_1, u_6\}, \{u_2\}, \{u_3\}, \{u_4\}, \{u_5\}\}$ will divide the objects based on their colour.

Given objects in U and an equivalence relation R_{colour} on U, concept X=*all blue and yellow objects* is R_{colour} definable while concept Y=*bright coloured objects* is not R_{colour} definable.

X is definable according to R_{colour} because it can be described precisely by the two subsets generated from R_{colour}, $\{u_2\}$ and $\{u_5\}$. Y is not definable because it cannot be derived in this way. Therefore, two subsets, one for lower approximation and another for upper approximation, are used to describe the possible objects that may suit the description given by Y. These two subsets are $\underline{(R_{colour})}Y = \{u_2\}$ and $\overline{(R_{colour})}Y = \{u_2, u_5\}$, if the colour of object u_5 'blue' is believed to be bright as well.

Rough sets theory has been successfully applied in many areas, such as in medical diagnosis, process control, knowledge discovery, and economics. Unlike other approaches to uncertainty reasoning which require additional model assumptions in information processing procedure, this theory solely uses the information provided by domain data to derive knowledge or useful information for decisions, as summarized by Pawlak [Pawlak, 1991]:

Table 1.4: A collection of objetcs

U	Colour	Shape	Size	Weight	Material
u_1	dark brown	square	small	heavy	copper
u_2	yellow	round	large	light	wood
u_3	black	square	small	heavy	steel
u_4	dark brown	triangular	large	light	plastic
u_5	blue	square	medium	light	cardboard
u_6	black	ellipse	small	heavy	lead

> "The numerical value of imprecision is not pre-assumed, as it is in probability theory or fuzzy sets - but is calculated on the basis of approximations which are the fundamental concepts used to express imprecision of knowledge ... As a result we do not require that an agent assigns precise numerical values to express imprecision of his knowledge, but instead imprecision is expressed by quantitative concepts (approximations)".

There has been, however, some works on integrating numerical measurement into rough sets theory. For example, it was proposed to use probabilities to indicate degrees of errors of resultant rules obtained using a rough set classifier ([Xiang *et al.*, 1993]), to create probabilistic rule induction ([Tsumoto and Tanaka, 1993], [Ziarko, 1993a]), to the development of probabilistic rough sets models ([Pawlak *et al.*, 1988], [Wong and Ziarko, 1987]). Also, an attempt to combine rough sets theory with Bayesian decision theory was investigated in ([Yao and Wong, 1992]).

We shall study rough sets theory in more detail in Chapter 10.

1.6 Incidence Calculus

What we have discussed so far is the study of integrating a numerical method with a symbolic one to obtain a more powerful mechanism for representing and reasoning with information. However, we are more interested in whether there is a single theory which possesses both numerical and symbolic approaches. The approach of this book is via incidence calculus.

Incidence calculus [Bundy, 1985] is a probabilistic logic developed from propositional logic by associating probabilities with formulae indirectly, in contrast to Nilsson's approach.

In incidence calculus, for a formula (sentence), denoted as ϕ, instead of saying ϕ is true or false, we say ϕ being true is supported by a set of possible worlds, and the probability of ϕ is defined as the sum of probabilities of these possible worlds.

If, for every sentence in incidence calculus, its probability is either 0 or 1, then the theory reduces to the traditional propositional logic case.

Consider the truth value of a sentence in a particular domain, denoted as $\mathcal{W}$, which contains possible worlds (called samples in probability theory) related to the sentence. Any event in $\mathcal{W}$ will either support the sentence (make it true) or refute the sentence (make it false). If we put all the possible worlds supporting the sentence together, called W_1, then we get a subset of $\mathcal{W}$. W_1 is called the *incidence set* of this sentence.

The probability of a sentence is defined as the probability of this subset. Therefore, to calculate the probability of a formula, we need to derive the incidence set of the formula first. All the formulae with initially known incidence sets are in a set, $\mathcal{A}$. The incidence set of any other formula not in $\mathcal{A}$ is derived through logical entailment.

For instance, if we know that the incidence sets of $q_1 \rightarrow q_2$ and $q_2 \rightarrow q_3$ are W_1 and W_2, meaning W_1 and W_2 support statements $q_1 \rightarrow q_2$ and $q_2 \rightarrow q_3$ respectively, then logically $W_1 \cap W_2$ should support $(q_1 \rightarrow q_2) \wedge (q_2 \rightarrow q_3)$, and further $q_1 \rightarrow q_3$. In this way, the incidence set of $q_1 \rightarrow q_3$ is changed from unknown to at least $W_1 \cap W_2$.

A set of possible worlds is located at the middle level in incidence calculus. It acts as a bridge between formulae and probabilities. Through this bridge, numerical values are assigned to hypotheses. This bridge, as we will see later, is very important in making the links between two different reasoning patterns. It is also because of this bridge, that incidence calculus is a good example of first type of hybrid systems where symbolic reasoning dominates the whole reasoning process.

1.7 Structure of the Book

The overall aim of this book is to demonstrate the importance and the possibility of performing symbolic and numerical reasoning in one structure, either through combining two separate mechanisms or employing one existing theory. Incidence calculus is chosen as the representative theory which unifies both symbolic and numerical reasoning features.

Incidence calculus is also compared with DS theory and the ATMS, the two representatives of purely numerical and symbolic reasoning approaches, in the context of representing and reasoning with uncertain and incomplete information.

Chapter 1 has analysed the strengths and limitations of both symbolic and numerical reasoning mechanisms. It has provided intensive reviews of some attempts at integrating these mechanisms to solve complex problems over the past 20 years.

Chapter 2 introduces incidence calculus in greater detail. This includes original definitions of incidence calculus theories, the Legal Assignment Finder for calculating lower and upper bounds of incidence and incidence assignment mechanisms.

Chapter 3 concentrates on how to generalize the original incidence calculus developed to a more general form. This is done by dropping some of the conditions on the incidence function, i, in the original incidence calculus. The generalized incidence calculus has the ability to represent ignorance. An important function, the *basic incidence assignment*, is defined, and the delicate relationships between the incidence function and the basic incidence assignment are discussed.

Chapter 4 presents a fast algorithm for incidence assignment under the concept of generalized incidence calculus. The distinctive nature of this algorithm is examined against a more general background of probability. The results in this chapter reveal the possibility of recovering a symbolic assignment from initial numerical assignments, with minimum effort.

Chapter 5 concentrates on the issue of combining multiple pieces of evidence, in particular, non-independent evidence. A combination mechanism is proposed in generalized incidence calculus which can combine both dependent and independent pieces of evidence. This new combination mechanism is compared with a number of well-known examples which are designed to show the limitations of DS theory.

Chapter 6 contains our contributions to the clarification of the problems with Dempster's combination rule in the context of *Dempster's original combination framework* [Dempster, 1967]. A few examples are used to demonstrate the significance of our contribution.

Some other aspects of DS theory are also reviewed briefly, such as its computational complexity and its difficulty in representing heuristic knowledge, as well as parallel techniques employed to speed up the evidence combination procedure.

Chapter 7 gives a comprehensive comparison between incidence calculus and DS theory. The results reveal that: (i) both theories have the same ability to represent evidence, (ii) they have the same ability to combine DS-independent evidence and achieve the same result, and (iii) incidence calculus can be taken as an alternative to DS theory in combining dependent evidence.

Chapter 8 reviews the ATMS [de Kleer, 1986a] and extends the original ATMS into a probabilistic oriented structure. This chapter lays the basis for the discussion in the next chapter.

Chapter 9 focuses on the relationship between generalized incidence calculus and the ATMS. Because of its symbolic nature, generalized incidence calculus is proved to be equivalent to the probabilistic based ATMS. In addition, generalized incidence calculus provides a basis for constructing probabilistic based ATMSs and supplying justifications for the ATMS automatically.

Chapter 10 first devotes its attention to rough sets theory. Its common components with generalized incidence calculus are identified. Then, generalized incidence calculus is compared with some related work which is not shown in detail in previous chapters. In the final section of this closing chapter, further work on incidence calculus is discussed, along with a short summary concluding the book.

Basically this book may be viewed to consist of two parts: the first, that includes Chapters 1–5, devoted to the development of incidence calculus itself, and the second, that includes Chapters 6–10, dealing with relationships between incidence calculus and other theories.

1.8 Remarks

In this chapter and later on throughout the book, words *uncertain(ty), incomplete(ness), imprecise(ion), vague(ness)* and *ignorance* are constantly being used to describe information.

Here we try to clarify what each of these words means in the context of this book. Parsons in [Parsons, 1996] provides a comprehensive survey of possible explanations of these words in terms of database and information systems. Here we summarize his main points.

Bonissone and Tong [Bonissone and Tong 1985] proposed one of the earliest classifications in which three types of imperfection in information were identified:

- *uncertainty* is related to the reliability of information, either as a result of misjudgements of observers or due to inaccuracy and poor reliability of the instruments used to make the observations,

- *incompleteness* arises from the absence of a value or lack of information, i.e., allowing *unknown* being the truth value of a statement, and

- *imprecision* occurs when a value cannot be measured with suitable precision.

Dubois and Prade in [Dubois and Prade,1988], as well as Bosc and Prade in [Bosc and Prade, 1993], voiced their general agreement with this classification but with the following modifications:

- *Uncertainty* may also arise due to lack of information. The result of lacking information makes it impossible to assess a statement (or fact) with accuracy and certainty. It is only possible to estimate the truth value of the statement with a suitable measurement.

- *Imprecision* is the result of unsuitable granularity in a measurement system.

- *Uncertainty and imprecision* are likely to appear in the same information simultaneously.

In this book, we prefer to follow Dubois and Prade's, and Bosc and Prade's interpretations of these definitions.

Vagueness is very much related to fuzzy values in fuzzy logic. Some descriptive words, such as, 'old', 'tall', 'very far', or 'very close' are usually involved in a vague value assigned to a statement. As we are not dealing with fuzzy logic in this book, the word *vagueness* here merely has the same meaning as *imprecision.*

1.9 Summary

Representing and reasoning with uncertain, imprecise and incomplete information has become one of the most important research areas in the artificial intelligence domain. The unsatisfactory performance of classical logic in these circumstances invoked studies of plausible reasoning both in pure symbolic form – non-monotonic reasoning and in numerical form – various numerical based reasoning mechanisms.

Classical logic was tailored to ideal situations where whatever is known is certain, and whatever is believed to be true is true forever.

Non-monotonic logics are designed to draw plausible conclusions from general rules when there are no contradictions, and to throw away no-longer-possible conclusions, in the light of new information. Conclusions in a non-monotonic logic do not need to be absolutely true, although some of them are. However, non-monotonic logics are not designed to show the degree of certainty of a conclusion, because they are purely symbolic and it is not possible to indicate an agent's degree of belief in a conclusion.

Numerical reasoning systems, on the other hand, are made to fit a human being's reasoning habit, in the situation where numbers indicates that a person is almost, but not absolutely sure, what she knows.

These two forms of plausible reasoning address different aspects of our common sense reasoning; one focuses on the soundness of the reasoning procedure, the other favours maximum accuracy of drawn conclusions. Combining the strengths of these two forms would certainly help to design a more powerful reasoning mechanism. That is what the many attempts mentioned above have been trying to achieve, and that is what we are aiming to study in this book.

Chapter 2

Incidence Calculus

Having discussed the advantages and disadvantages of pure numerical and symbolic approaches to reasoning under uncertainty, and a number of attempts at integrating these two reasoning mechanisms, we are ready to study incidence calculus in detail.

In this chapter, we introduce the original incidence calculus developed by Bundy [Bundy, 1985]. We discuss its main features, its *Legal Assignment Finder* for deriving lower and upper bounds of incidences, and two methods for assigning incidences based on numerical assignments. Some limitations of the original incidence calculus are also investigated.

2.1 Incidence Calculus Theories

Incidence calculus ([Bundy, 1985], [Bundy, 1992]) is a probabilistic logic for dealing with uncertainty in intelligent systems, through extending classical propositional logic. It was initially developed to overcome the problem of dealing with, interpreting, and propagating numerical values in pure numerical uncertainty reasoning techniques.

In incidence calculus, incidences are assigned to formulae, and probabilities are assigned to incidences. An incidence is assigned to a formula if the formula is true when the incidence occurs. Incidences can be explained as either the possible worlds relevant to a problem in logic, or the possible outcomes of an event in probability theory. The probability of a formula is calculated through the set of incidences assigned to it.

2.1.1 Basic Definitions

Propositional logic

The symbols of propositional logic are atomic propositional symbols (variables):

$$q, q_1, q_2, \ldots,$$

truth symbols (logical constants):

$$true, false,$$

logical connectives:

$$\neg, \wedge, \vee, \rightarrow, \leftrightarrow,$$

and parentheses

$$(,).$$

Propositional symbols, denote *atomic propositions* or *statements*, that may be either true or false, such as "it is raining" or "the car is white".

An atomic proposition is a proposition which cannot be further broken down into simpler propositions (statements).

Formulae (or sentences) in propositional logic are formed from these atomic symbols based on the following rules:

- Every atomic proposition and truth symbol is a sentence.
- The negation ($\neg$) of a sentence is a sentence, such as $\neg q$ from q.
- The conjunction ($\wedge$) of two sentences is a sentence, such as $q_1 \wedge q_2$.
- The disjunction ($\vee$) of two sentences is a sentence, such as $q_1 \vee q_2$.
- The implication ($\rightarrow$) of one sentence for another is a sentence, such as $q_1 \rightarrow q_2$.
- The equivalence of two sentences is a sentence, such as $q_1 \leftrightarrow q_2$.

For instance, $q \rightarrow (q_1 \wedge q_2)$ is a formula (sentence), so is $(q \rightarrow q_1) \rightarrow (q_2 \vee q_3)$. Parentheses can be dropped if we assign these five logical connectives in a order and require that the connective with lower rank applies first. Conventionally, these five connectives are given ascending orders

$$\neg, \wedge, \vee, \rightarrow, \leftrightarrow .$$

In this way, formula $(q \rightarrow q_1) \rightarrow (q_2 \vee q_3)$ can be rewritten as $(q \rightarrow q_1) \rightarrow q_2 \vee q_3$ without any confusion.

A sentence (or formula) may be either true or false, but not both, given some state of the world. The truth value assignment to a sentence is called an *interpretation*, an assertion about its truth in some possible

world. For example, the proposition "the car is white" is *true* when an agent is referring to a white car and is *false* when the car is red.

Formally, an interpretation of a sentence is a mapping from propositional symbols into the set $\{true, false\}$ which can be shown as a row in a *truth*

Table 2.1: Truth table of five logical connectives

q_1	q_2	$\neg q_1$	$q_1 \wedge q_2$	$q_1 \vee q_2$	$q_1 \to q_2$	$q_1 \leftrightarrow q_2$
true	*true*	*false*	*true*	*true*	*true*	*true*
true	*false*	*false*	*false*	*true*	*false*	*false*
false	*true*	*true*	*false*	*true*	*true*	*false*
false	*false*	*true*	*false*	*false*	*true*	*true*

table. A truth table takes truth values of propositional symbols as input and returns truth values of a sentence as output. Table 2.1 outlines the truth values of formulae $\neg q_1$, $q_1 \wedge q_2$, $q_1 \vee q_2$, $q_1 \to q_2$ and $q_1 \leftrightarrow q_2$ in relation to the truth values of atomic propositions q_1 and q_2.

For example, the mapping (q_1, q_2) to $(false, true)$ respectively is an interpretation of formula $q_1 \wedge q_2$, as well as $q_1 \to q_2$, where $q_1 \wedge q_2$ has truth value *false* and $q_1 \to q_2$ has truth value *true*. This interpretation is reflected in the third row of Table 2.1.

The truth assignments of compound propositions (formulae) can be calculated solely from their constituent parts. For instance, the truth value of $(q_1 \vee q_1) \wedge \neg q_2$ is *true* if and only if q_1 has truth value *true* and q_2 has *false*. This feature is called *truth functional*. For logical constants, the truth value of sentence *true* is always *true* and the truth value of sentence *false* is always *false*.

A formula ϕ is said to be *valid* or a *tautology* if ϕ has truth value *true* under all possible interpretations and is named *invalid* if it has truth value *false* in all interpretations.

In the following, we use *formula* (or *sentence*) to name either an atomic proposition or a compound statement.

Given two formulae ϕ and ψ, notation $\psi \models \phi$ is used when $\psi \to \phi$ is a tautology (valid) and notation $\psi = \phi$ means that sentence $\psi \leftrightarrow \phi$ is always true, that is, $\psi \models \phi$ and $\phi \models \psi$,indicating that ψ and ϕ are semantically equivalent.

Possible worlds

Each possible world is a primitive object which can be thought of as a partial interpretation of some logical formulae or taken as a possible answer to a particular question.

The probability is represented by a function $\mu : \mathcal{W} \to [0,1]$, where $\mathcal{W}$ is a finite set of possible worlds[1]. For each possible world $w \in \mathcal{W}$, $\mu(w)$ is known and the probability sum of the whole set $\mathcal{W}$ is 1, that is $\mu(\mathcal{W}) = 1$.

For example, consider the situation where a new module 'Introduction to Artificial Intelligence' is to be allocated to a lecturer. There are ten lecturers,

[1]This requirement will not affect the discussion in the rest of the book.

of whom six are not suitable, and the remaining four are equally probable to teach the module. Therefore, there are ten possible worlds in which the AI module might be taught, with probability distribution $\mu(q_j) = 0$ for $j = 1, \ldots, 6$ and $\mu(q_j) = 1/4$ for $j = 7, \ldots, 10$.

Definition 2.1: *Propositional Language $\mathcal{L}(P)$*

- *P is a finite set of atomic propositions.*
- *$\mathcal{L}(P)$ is the propositional language formed from P:*
  $true, false \in \mathcal{L}(P)$,
  if $q \in P$, then $q \in \mathcal{L}(P)$, and
  if $\phi, \psi \in \mathcal{L}(P)$ then $\neg\phi, \neg\psi \in \mathcal{L}(P)$,
  $\phi \wedge \psi \in \mathcal{L}(P)$, $\phi \vee \psi \in \mathcal{L}(P)$, and
  $\phi \rightarrow \psi, \psi \rightarrow \phi \in \mathcal{L}(P)$.
  That is, $\mathcal{L}(P)$ is closed under the logical connectives negation ($\neg$), disjunction ($\vee$), conjunction ($\wedge$) and implication ($\rightarrow$).

Definition 2.2: *Basic Element Set*

Assume P is a finite set of propositions with $P = \{q_1, q_2, \ldots, q_n\}$.

An item δ, defined as $\delta = q'_1 \wedge \ldots \wedge q'_n$ where q'_j is either q_j or $\neg q_j$, is called a basic element.

The collection of all the basic elements, denoted as $\mathcal{A}t$, is called the basic element set of P. Any formula ψ in the language set $\mathcal{L}(P)$ can be represented as

$$\psi = \delta_1 \vee \ldots \vee \delta_k,$$

where $\delta_j \in \mathcal{A}t$.

Definition 2.3: *Probability Space*

A probability space (X, χ, μ) has:

- *X: a sample space;*
- *χ: a σ-algebra containing some subsets of X, which is defined as containing X and closed under complementation and countable union.*
- *μ: a probability measure $\mu : \chi \rightarrow [0, 1]$ with the following features:*

  P1. *$\mu(X_i) \geq 0$ for all $X_i \in \chi$;*

  P2. *$\mu(X) = 1$;*

P3. $\mu(\cup_{j=1}^{\infty} X_j) = \Sigma_{j=1}^{\infty} \mu(X_j)$, *if the* X_j*'s are pairwise disjoint members of* χ.

A subset χ' of χ is called a *basis* of χ if it contains non-empty and disjoint elements, and if χ consists precisely of countable unions of members of χ'. For any finite χ there is a unique basis χ' of χ and it follows that

$$\Sigma_{X_i \in \chi'} \mu(X_i) = 1.$$

Given a set X, when a probability distribution μ assigns a probability on every singleton $x \in X$, σ-algebra χ of X is 2^X and the basis χ' is $\{\{x\} \mid x \in X\}$. The corresponding probability space is $(X, 2^X, \mu)$.

2.1.2 Incidence Calculus Theories

In incidence calculus, the mechanism for representing knowledge and evidence is the incidence calculus theory, in which a set of axioms are assigned with incidences initially. Incidence calculus theories are the bodies for carrying evidence and knowledge, and the sources for deriving conclusions.

Definition 2.4: *Incidence Calculus Theory*

An incidence calculus theory is a quintuple

$$< \mathcal{W}, \mu, P, \mathcal{A}, i >,$$

where:

- $\mathcal{W}$ *is a finite set of possible worlds.*
- *For all* $w \in \mathcal{W}$, $\mu(w)$ *is the probability of* w *and* $\mu(\mathcal{W}) = 1$, *where* $\mu(I) = \Sigma_{w \in I} \mu(w)$.
- P *is a finite set of atomic propositions.* $\mathcal{A}t$ *is the basic element set of* P. $\mathcal{L}(P)$ *is the language set of* P. *For any atomic proposition* $q \in P$, q *may have either true or* $false$ *(not both) as its truth value.*
- $\mathcal{A}$ *is a distinguished set of formulae in* $\mathcal{L}(P)$ *called the axioms of the theory.*
- i *is a function from the axioms in* $\mathcal{A}$ *to* $2^{\mathcal{W}}$, *the set of subsets of* $\mathcal{W}$. $i(\phi)$ *is to be thought of as the set of possible worlds in* $\mathcal{W}$ *in which* ϕ *is true, i.e.,* $i(\phi) = \{w \in \mathcal{W} \mid w \models \phi\}$. $i(\phi)$ *is called the incidence set of* ϕ.

i can be extended as a function from $\mathcal{L}(\mathcal{A})$ to $2^{\mathcal{W}}$ using the following defining equations of incidences:

$$\begin{aligned} i(true) &= \mathcal{W}, \\ i(false) &= \emptyset, \\ i(\neg\phi) &= \mathcal{W} \setminus i(\phi), \\ i(\phi \wedge \psi) &= i(\phi) \cap i(\psi), \\ i(\phi \vee \psi) &= i(\phi) \cup i(\psi), \\ i(\phi \to \psi) &= \mathcal{W} \setminus i(\phi) \cup i(\psi). \end{aligned}$$

In addition, when $\phi = \psi$ and $i(\phi)$ is given, $i(\psi)$ is defined as $i(\phi)$.

Here, the notation $w \models \phi$ is explained as ϕ is true under the interpretation of w, and $\emptyset$ stands for the empty set. Such an incidence calculus theory is *truth functional.*

Because an incidence function i can be extended as a function on $\mathcal{L}(\mathcal{A})$ on which it should be truth functional, it is expected that i should be truth functional on $\mathcal{A}$ as well.

Given an incidence calculus theory, it should be verified first whether i satisfies this requirement on $\mathcal{A}$. For instance, if formula ϕ can be expressed as

$$\phi = (\psi_1 \vee \psi_2) = (\psi_3 \vee \psi_4),$$

and $i(\psi_j), j = 1, \ldots, 4$ are all known, then $i(\psi_1) \cup i(\psi_2)$ must be the same as $i(\psi_3) \cup i(\psi_4)$; otherwise this initial incidence assignment is not consistent.

For a formula in $\mathcal{L}(P) \setminus \mathcal{L}(A)$, it is not possible to obtain its incidence set. Instead, it is only sensible to infer the lower and upper bounds of an incidence set. These bounds are defined as [Bundy, 1985]:

$$i_*(\phi) = \bigcup_{\psi \in \mathcal{L}(A)} \{i(\psi) \mid i(\psi \to \phi) = \mathcal{W}\}; \tag{2.1}$$

and

$$i^*(\phi) = \bigcap_{\psi \in \mathcal{L}(A)} \{i(\psi) \mid i(\phi \to \psi) = \mathcal{W}\}. \tag{2.2}$$

Notation $i(\psi \to \phi) = \mathcal{W}$ implies that the truth value of formula $\psi \to \phi$ is *true*. Alternatively, $i(\psi \to \phi) = \mathcal{W}$ can be restated as $\psi \models \phi$.

For the upper and lower bounds of a formula, we have the following two propositions [Correa da Silva and Bundy, 1991].

Proposition 2.1: $\forall \phi \in \mathcal{L}(\mathcal{A}),\ i_*(\phi) = i(\phi) = i^*(\phi)$.

Proof.

Suppose that $i_*(\phi) \neq i(\phi)$. Then, either $i_*(\phi) \not\subseteq i(\phi)$ or $i_*(\phi) \not\supseteq i(\phi)$. But

$$
\begin{aligned}
& i_*(\phi) \not\subseteq i(\phi) \\
& \Rightarrow \exists \psi \in \mathcal{L}(\mathcal{A}):\ i(\psi) \not\subseteq i(\phi), i(\psi \to \phi) = \mathcal{W} \\
& \Rightarrow i(\neg(\psi \wedge (\neg\phi))) = \mathcal{W} \setminus (i(\psi) \cap i(\neg\phi)) = \mathcal{W} \\
& \Rightarrow i(\psi) \cap i(\neg\phi) = \emptyset,
\end{aligned}
$$

which contradicts the hypothesis.

The non-logical notation $\Rightarrow$ is explained as: $a \Rightarrow b$ means if the antecedent a is true, then the consequent b is true as well.

On the other hand,

$$
i_*(\phi) \not\supseteq i(\phi)
$$

$$
\Rightarrow \exists \psi = \phi \wedge \neg[\vee\phi_j : \phi_j \in \mathcal{L}(\mathcal{A}), \phi_j \models \phi] \in \mathcal{L}(\mathcal{A}) : i(\psi) \subseteq i(\phi), i(\psi) \not\subseteq i_*(\phi),
$$

which contradicts the hypothesis since $i(\psi \to \phi) = \mathcal{W}$.

Now, suppose that $i^*(\phi) \neq i(\phi)$. Then, either $i^*(\phi) \not\supseteq i(\phi)$ or $i^*(\phi) \not\subseteq i(\phi)$. But

$$
\begin{aligned}
& i^*(\phi) \not\supseteq i(\phi) \\
& \Rightarrow \exists \psi \in \mathcal{L}(\mathcal{A}):\ i(\psi) \not\supseteq i(\phi), i(\phi \to \psi) = \mathcal{W} \\
& \Rightarrow i(\neg(\phi \wedge (\neg\psi))) = \mathcal{W} \setminus (i(\phi) \cap i(\neg\psi)) = \mathcal{W} \\
& \Rightarrow i(\phi) \cap i(\neg\psi) = \emptyset,
\end{aligned}
$$

which contradicts the hypothesis.

On the other hand,

$$
i^*(\phi) \not\subseteq i(\phi)
$$

$$
\Rightarrow \exists \psi = \neg\phi \wedge [\wedge\phi_j : \phi_j \in \mathcal{L}(\mathcal{A}), \phi \models \phi_j] \in \mathcal{L}(\mathcal{A}) : i(\psi) \subseteq i_*(\phi), i(\psi) \not\subseteq i(\phi),
$$

which contradicts the hypothesis since $i(\psi \to \neg\phi) = \mathcal{W}$.

QED

Proposition 2.2: $\forall \phi \in \mathcal{L}(P),\ i_*(\phi) = \mathcal{W} \setminus i^*(\neg\phi)$.

Proof.

$$
\begin{aligned}
\mathcal{W} \setminus i^*(\neg\phi) &= \mathcal{W} \setminus \bigcap[i(\psi) : \psi \in \mathcal{L}(\mathcal{A}), i(\neg\phi \to \psi) = \mathcal{W}] \\
&= \bigcup[\mathcal{W} \setminus i(\psi) : \psi \in \mathcal{L}(\mathcal{A}), i(\neg\phi \to \psi) = \mathcal{W}] \\
&= \bigcup[i(\neg\psi) : \psi \in \mathcal{L}(\mathcal{A}), i(\neg\psi \to \phi) = \mathcal{W}] \\
&\qquad\qquad \text{(applying } i(\neg\psi) = \mathcal{W} \setminus i(\psi)\text{)} \\
&= i_*(\phi).
\end{aligned}
$$

QED

Proposition 2.1 says that when ϕ is in $\mathcal{L}(\mathcal{A})$, its lower and upper bounds are the same. Proposition 2.2 states that when $\phi \notin \mathcal{L}(\mathcal{A})$, $i_*(\phi) \neq i^*(\phi)$. In fact, in this case, $i_*(\phi) \subseteq i^*(\phi)$ ([McLean, 1992]).

The corresponding lower and upper bounds of probabilities of formula ϕ are $prob_*(\phi) = \mu(i_*(\phi))$ and $Prob^*(\phi) = \mu(i^*(\phi))$. For a formula, if $i_*(\phi) = i^*(\phi) = i(\phi)$, then $p(\phi)$ is defined as $Prob_*(\phi)$ which is referred to as the probability of formula ϕ. Therefore, for every formula in $\mathcal{L}(\mathcal{A})$, $p(\phi)$ is known and $p(\phi) = \mu(i(\phi))$.

For any two formulae ϕ and ψ in $\mathcal{L}(\mathcal{A})$, the conditional probability of ϕ given ψ is defined as

$$p(\phi \mid \psi) = \frac{p(\phi \wedge \psi)}{p(\psi)}. \tag{2.3}$$

Example 2.1

Suppose there are three propositions, $P = \{sunny, rainy, windy\}$, and seven possible worlds, $\mathcal{W} = \{sun, mon, tues, wed, thus, fri, sat\}$. Assume that each possible world is equally probable.

Through a piece of evidence, we learn that four possible worlds *fri, sat, sun, mon* support *rainy*, and three possible worlds *mon, wed, fri* make *windy* true.

Then, the incidence sets of these two propositions are:

$$\begin{aligned} i(rainy) &= \{fri, sat, sun, mon\}, \\ i(windy) &= \{mon, wed, fri\}. \end{aligned}$$

The set of axioms $\mathcal{A}$ is $\{rainy, windy\}$. For every formula in $\mathcal{L}(\mathcal{A})$, it is possible to get its incidence set. For instance, for formula $\neg windy \wedge rainy$, its incidence set is

$$\begin{aligned} i(\neg windy \wedge rainy)\ & i(\neg windy) \cap i(rainy) \\ &= (\mathcal{W} \setminus i(windy)) \cap i(rainy) \\ &= \{tue, thru, sat, sun\} \cap \{fri, sat, sun, mon\} \\ &= \{sat, sun\}. \end{aligned}$$

So that $i_*(\neg windy \wedge rainy) = i(\neg windy \wedge rainy) = i^*(\neg windy \wedge rainy)$.

For other formulae in $\mathcal{L}(P) \setminus \mathcal{L}(A)$, such as, $\phi = (sunny \vee windy) \wedge \neg rainy$, equations (2.1) and (2.2) are used to obtain their bounds.

In this special case, $i_*(\phi) = i(windy \wedge \neg rainy) = \{wed\}$ (as only $i((windy \wedge \neg rainy) \rightarrow \phi) = \mathcal{W}$) and $i^*(\phi) = \{tues, wed, thus\}$ (as $i(\phi \rightarrow \neg rainy) = \mathcal{W}$).

The result says that there is at least one day, but at most three days in the week that are either sunny or windy but not rainy. [2]

2.2 The Legal Assignment Finder

When an incidence calculus theory is specified, it is guaranteed that the known evidence gives a precise incidence set and a probability on every formula in $\mathcal{L}(\mathcal{A})$ as well as lower and upper bounds of incidence sets and probabilities on other formulae. The obvious method of deriving these bounds are the direct application of equations (2.1) and (2.2). However, this method is inefficient when $\mathcal{L}(\mathcal{A})$ is large as great effort is needed to carry out tautology checking.

On the other hand, it may also be the case that a piece of evidence can only specify the lower and upper bounds of incidences on some formulae without giving the incidence function explicitly [Wong et al, 1992].

Therefore, an inference mechanism is required to infer bounds for other formulae based on these given bounds. Assume that the initial lower and upper bounds of incidences on a formula ϕ in a subset S of $\mathcal{L}(P)$ are $inf(\phi)$ and $sup(\phi)$ respectively[3].

Then, (inf, sup) can be extended as two mappings from $\mathcal{L}(P)$ to $2^{\mathcal{W}}$ by assigning $inf(\psi) = \emptyset$ and $sup(\psi) = \mathcal{W}$ for $\psi \in \mathcal{L}(P) \backslash S$. The lower and upper bounds of incidences on a formula given in this way are not tight enough.

The approach for finding tighter bounds through the initial (inf, sup) is called the *Legal Assignment Finder* [Bundy, 1985] [Bundy, 1986].

2.2.1 Assignments and Inference Rules

Definition 2.5: *Assignments*

An assignment G consists of a pair of mappings (inf_G, sup_G) from a subset S of $\mathcal{L}(P)$ to $2^{\mathcal{W}}$. inf_G is called the lower incidence bound and sup_G the upper incidence bound of formula ϕ in S.

inf_G and sup_G can be extended as mappings from $\mathcal{L}(\mathcal{P})$ to $2^{\mathcal{W}}$ by assigning $ing_G(\psi) = \emptyset$ and $sup_G(\psi) = \mathcal{W}$ as default when $\psi \notin S$.

[2] Symbol ◇ is used to indicate the end of an example through out the book.

[3] In fact, an incidence function i defined on $\mathcal{L}(\mathcal{A})$ is a special case of this kind, if we let $S = \mathcal{L}(\mathcal{A})$ and $inf(\phi) = i_*(\phi)$ and $sup(\phi) = i^*(\phi)$.

Definition 2.6: *Inference Rules*

An inference rule is a mapping from one assignment to another. If G_1 is the assignment before a rule is applied and G_2 is the assignment after the rule is applied, then this rule changes the assignment from G_1 status to G_2 status.

Obtaining an assignment G_2 from G_1, through firing a rule may not change any bound in assignment G_1, that is, for all $\phi \in S$, $inf_{G_2}(\phi) = inf_{G_1}(\phi)$ and $sup_{G_2}(\phi) = sup_{G_1}(\phi)$. In fact, this rule may have changed nothing. This is not the purpose of applying rules on assignments at all. What matters is that if a rule fires, there is at least one formula ϕ whose lower bound or upper bound is changed, so that a tighter bound results.

Definition 2.7: *Derived Assignments*

Let G_1 be an assignment consisting of a pair (inf_{G_1}, sup_{G_1}) from S to $2^{\mathcal{W}}$. If a pair of inference rules (r_1, r_2) could either produce $inf_{G_2}(\phi) \neq inf_{G_1}(\phi)$ or $sup_{G_2}(\phi) \neq sup_{G_1}(\phi)$ or both for a specific ϕ, then G_2 is called a derived assignment from G_1 through (r_1, r_2). For other $\psi \in S$, define $inf_{G_2}(\psi) = inf_{G_1}(\psi)$ and $sup_{G_2}(\psi) = sup_{G_1}(\psi)$. G_1 is named as a base assignment of G_2.

The following basic inference rules for negation and conjunction were developed by Bundy [Bundy, 1985]:

$$
\begin{array}{lrcl}
\text{Not}_1: & sup_{G_2}(\phi) & = & (\mathcal{W} \setminus inf_{G_1}(\neg\phi)) \cap sup_{G_1}(\phi) \\
\text{Not}_2: & inf_{G_2}(\phi) & = & (\mathcal{W} \setminus sup_{G_1}(\neg\phi)) \cup inf_{G_1}(\phi) \\
\text{Not}_3: & sup_{G_2}(\neg\phi) & = & (\mathcal{W} \setminus inf_{G_1}(\phi)) \cap sup_{G_1}(\neg\phi) \\
\text{Not}_4: & inf_{G_2}(\neg\phi) & = & (\mathcal{W} \setminus sup_{G_1}(\phi)) \cup inf_{G_1}(\neg\phi) \\
\text{And}_1: & sup_{G_2}(\phi) & = & (sup_{G_1}(\phi \wedge \psi) \cup (\mathcal{W} \setminus inf_{G_1}(\psi))) \cap sup_{G_1}(\phi) \\
\text{And}_2: & inf_{G_2}(\phi) & = & inf_{G_1}(\phi \wedge \psi) \cup inf_{G_1}(\phi) \\
\text{And}_3: & sup_{G_2}(\psi) & = & (sup_{G_1}(\phi \wedge \psi) \cup (\mathcal{W} \setminus inf_{G_1}(\phi))) \cap sup_{G_1}(\psi) \\
\text{And}_4: & inf_{G_2}(\psi) & = & inf_{G_1}(\phi \wedge \psi) \cup inf_{G_1}(\psi) \\
\text{And}_5: & sup_{G_2}(\phi \wedge \psi) & = & sup_{G_1}(\phi) \cap sup_{G_1}(\psi) \cap sup_{G_1}(\phi \wedge \psi) \\
\text{And}_6: & inf_{G_2}(\phi \wedge \psi) & = & (inf_{G_1}(\phi) \cap inf_{G_1}(\psi)) \cup inf_{G_1}(\phi \wedge \psi)
\end{array}
$$

To apply these rules to obtain new assignments, each formula has to be converted into a form which involves only negation and conjunction.

Definition 2.8: *Canonical Form* ([Bundy, 1992])

A formula ϕ in $\mathcal{L}(P)$ is in canonical form if it has the form $\bigwedge_{j=1}^{m}(\neg \bigwedge_{l=1}^{n} r_j^l)$ where $r_j^l = q$ or $r_j^l = \neg q$ for $q \in P$.

For each formula ϕ, it is always possible to transform it into a canonical form using the following three steps:

1. transform ϕ into a conjunctive normal form;

2. apply de Morgan's law to turn each conjunct $\bigvee_{l=1}^{m} r_j^l$ into $\neg \bigwedge_{l=1}^{m} \neg r_j^l$;

3. and, finally, cancel all double negations.

McLean ([McLean, 1992]) discovered more inference rules for the disjunction and implication:

$$\begin{array}{lrcl}
\text{Or}_1: & inf_{G_2}(\phi) & = & (inf_{G_1}(\phi \vee \psi) \cap (\mathcal{W} \setminus sup_{G_1}(\psi))) \cup inf_{G_1}(\phi) \\
\text{Or}_2: & sup_{G_2}(\phi) & = & sup_{G_1}(\phi \vee \psi) \cap sup_{G_1}(\phi) \\
\text{Or}_3: & inf_{G_2}(\psi) & = & (inf_{G_1}(\phi \vee \psi) \cap (\mathcal{W} \setminus sup_{G_1}(\phi))) \cup inf_{G_1}(\psi) \\
\text{Or}_4: & sup_{G_2}(\psi) & = & sup_{G_1}(\phi \vee \psi) \cap sup_{G_1}(\psi) \\
\text{Or}_5: & inf_{G_2}(\phi \vee \psi) & = & inf_{G_1}(\phi) \cup inf_{G_1}(\psi) \cup inf_{G_1}(\phi \vee \psi) \\
\text{Or}_6: & sup_{G_2}(\phi \vee \psi) & = & (sup_{G_1}(\phi) \cup sup_{G_1}(\psi)) \cap sup_{G_1}(\phi \vee \psi) \\
\text{Imp}_1: & sup_{G_2}(\phi) & = & (\mathcal{W} \setminus inf_{G_1}(\phi \rightarrow \psi) \cup sup_{G_1}(\psi)) \cap sup_{G_1}(\phi) \\
\text{Imp}_2: & inf_{G_2}(\phi) & = & (\mathcal{W} \setminus sup_{G_1}(\phi \rightarrow \psi)) \cup inf_{G_1}(\phi) \\
\text{Imp}_3: & sup_{G_2}(\psi) & = & sup_{G_1}(\phi \rightarrow \psi) \cap sup_{G_1}(\psi) \\
\text{Imp}_4: & inf_{G_2}(\psi) & = & (inf_{G_1}(\phi \rightarrow \psi) \cap inf_{G_1}(\phi)) \cup inf_{G_1}(\psi) \\
\text{Imp}_5: & sup_{G_2}(\phi \rightarrow \psi) & = & ((\mathcal{W} \setminus inf_{G_1}(\phi)) \cup sup_{G_1}(\psi)) \cap sup_{G_1}(\phi \rightarrow \psi) \\
\text{Imp}_6: & inf_{G_2}(\phi \rightarrow \psi) & = & (\mathcal{W} \setminus sup_{G_1}(\phi)) \cup inf_{G_1}(\psi) \cup inf_{G_1}(\phi \rightarrow \psi)
\end{array}$$

Theoretically, the inference rules for negation and conjunction are sufficient for deriving incidences since every formula in $\mathcal{L}(P)$ is semantically equivalent to one containing only $\neg$ and $\wedge$. However, using the disjunction and implication rules avoids the need to reduce formulae to expressions involving only $\neg$ and $\wedge$, and the additional connectives can make formulae easier to read.

It is proved in [Bundy, 1986] that the exhaustive application of negation and conjunction rules will terminate, and the inference results of using these rules are sound. In [Wang *et al.* 1996], it is proved that Bundy's rules are complete as well. It is also proved in [Correa da Silva and Bundy, 1991]that the final assignment (inf_F, sup_F) has the following properties:

$$inf_F(\phi) \subseteq i_*(\phi),$$

$$sup_F(\phi) \supseteq i^*(\phi),$$

$$lim_{F\rightarrow\infty} inf_F(\phi) = i_*(\phi),$$

and

$$lim_{F\rightarrow\infty} sup_F(\phi) = i^*(\phi).$$

2.2.2 Constraint Sets

Assume that the set of axioms to which an initial pair of lower and upper bounds is assigned is $\mathcal{A}$ and the set of formulae of particular interest is $\mathcal{A}'$. Usually the assignment of initial lower and upper bounds needs to be extended to a certain set in order to use the Legal Assignment Finder. This set is called the *constraint set* and is defined as follows.

Definition 2.9: *Constraint Set* ([Bundy, 1992])

Let $\mathcal{A}$ be the set of axioms of a theory and $\mathcal{A}'$ be the set of formulae whose bounds of incidences we are interested in calculating.

Let $sf(S)$ be the set of subformulae of the formulae S, i.e.

$$\begin{array}{l} \textit{if } \phi \in S \textit{ then } \phi \in sf(S); \\ \textit{if } \neg\phi \in sf(S) \textit{ then } \phi \in sf(S); \\ \textit{if } \phi \wedge \psi \in sf(S) \textit{ then } \phi, \psi \in sf(S); \\ \textit{if } \phi \vee \psi \in sf(S) \textit{ then } \phi, \psi \in sf(S); \\ \textit{if } \phi \rightarrow \psi \in sf(S) \textit{ then } \phi, \psi \in sf(S). \end{array}$$

The constraint set is $sf(\mathcal{L}(\mathcal{A} \cup \mathcal{A}'))$.

So, given an initial pair of assignments inf and sup, these assignments only need to be extended to formulae in $sf(\mathcal{L}(\mathcal{A} \cup \mathcal{A}'))$ to obtain the bounds of any formula in $\mathcal{A}'$.

2.2.3 Termination Decisions of the Inference Procedure

The Legal Assignment Finder is applicable in two situations, either an incidence function i is given on a set of axioms, $\mathcal{A}$, or a pair of mapping functions (inf, sup) are defined on a set of formulae, $\mathcal{A}$, and the interest is the bounds of incidences of some formulae in $\mathcal{L}(P) \setminus \mathcal{L}(\mathcal{A})$. In either case, a constraint set S through Definition 2.9 is necessary for further inference.

A set of inference rules proposed in [Bundy, 1985] and [McLean, 1992] can be applied repeatedly in the following *Incidence Calculus Inference Procedure* ([Correa da Silva and Bundy, 1991]).

Suppose that G is the assignment after some rules are fired and $inf_G(\phi)$ and $sup_G(\phi)$ are the lower and upper bounds assigned to ϕ. Then:

1. If there exists a formula ψ where $inf_G(\psi) \not\subseteq sup_G(\psi)$, then the initial incidence assignment is not consistent. Terminate the inference procedure.

2. Apply the first pair of rules which makes $inf_{G+1}(\phi) \neq inf_G(\phi)$ and $sup_{G+1}(\phi) \neq sup_G(\phi)$.

3. If there are only rules which make $inf_{G+1}(\phi) \neq inf_G(\phi)$ but none for $sup_{G+1}(\phi) \neq sup_G(\phi)$, then apply one of these rules and define $sup_{G+1}(\phi) = sup_G(\phi)$.

4. If there are only rules which make $sup_{G+1}(\phi) \neq sup_G(\phi)$ but none for $inf_{G+1}(\phi) \neq inf_G(\phi)$, then apply one of these rules and define $inf_{G+1}(\phi) = inf_G(\phi)$.

5. If there are no rules which make $sup_{G+1}(\phi) \neq sup_G(\phi)$ or $inf_{G+1}(\phi) \neq inf_G(\phi)$, then terminate the procedure.

It is proved in [Bundy, 1986] that the application of the inference rules in the above procedure to an initial incidence assignment (either a single i or a pair of bounds) will eventually terminate. This procedure may or may not produce a consistent incidence assignment, depending on the point at which the procedure terminates.

Bundy (cf. [Bundy, 1985] and [Bundy, 1986]) designed, proved and implemented a procedure called a *case splitting* to continue the search for a consistent assignment when the above procedure stops at Step 5.

It is said in [Bundy, 1985] that:

> "...case splitting is necessary when the inference process runs out of rules to fire without specializing the assignment to a total or contradictory one. It is done by picking a point (meaning a possible world) and considering the other cases that it is not in (step 1) and that it is in (step 2, 3, 4) the incidence of a sentence".

However, it is suggested in [Wang *et al.* 1996] that case splitting is not necessary.

2.3 Examples of Using the Legal Assignment Finder

We will show in this section how to use the inference rules introduced above to derive lower and upper incidence bounds of formulae.

In the first example, incidence function i on a subset of $\mathcal{L}(P)$ is specified, so the lower and upper bounds of the incidence of a formula can be obtained by either using equations (2.1) and (2.2) or using the Legal Assignment Finder.

In the second example, only the lower and upper bounds of incidences on formulae in a subset $\mathcal{A}$ of $\mathcal{L}(P)$ are given, so the Legal Assignment Finder is the only choice for inferring bounds.

Example 2.2

Let the set of axioms for a given incidence calculus theory be $\mathcal{A} = \{\phi, \phi \rightarrow \psi\}$, the set of possible worlds be $\mathcal{W} = \{a, b\}$. Let $i(\phi) = \{a\}$, $i(\phi \rightarrow \psi) = \{a, b\}$.

The incidence set of formula ψ in $\mathcal{A}' = \{\psi\}$ is an agent's interest.

The incidence set of ψ cannot be obtained directly as $\psi \notin \mathcal{L}(\mathcal{A})$, where $\mathcal{L}(\mathcal{A}) = \{\phi, \neg\phi, \phi \rightarrow \psi, \neg(\phi \rightarrow \psi), \phi \wedge \psi, \neg(\phi \wedge \psi)\}$ with additional incidence sets on formulae in $\mathcal{L}(\mathcal{A}) \setminus \mathcal{A}$ as:

$$i(\neg\phi) = \{b\}, \qquad i(\neg(\phi \rightarrow \psi)) = \emptyset,$$
$$i(\phi \wedge \psi) = \{a\}, \qquad i(\neg(\phi \wedge \psi)) = \{b\}.$$

There are two approaches to obtaining the bounds of incidence set of ψ, using equations (2.1) and (2.2) or applying the Legal Assignment Finder.

Approach 1: Using the equations for lower and upper bounds

As $i((\phi \wedge \psi) \rightarrow \psi) = \mathcal{W}$, the application of equation (2.1) to ψ gives

$$i_*(\psi) = i(\phi \wedge \psi) = \{a\}.$$

Similarly,

$$i^*(\psi) = i(\phi \rightarrow \psi) = \{a, b\}$$

by applying equation (2.2) on formula ψ, as $i(\psi \rightarrow (\phi \rightarrow \psi)) = \mathcal{W}$.

Therefore, the incidence set of ψ lies between $\{a\}$ and $\{a, b\}$.

Approach 2: Using the Legal Assignment Finder

Firing each inference rule normally leads from one assignment to another. This procedure terminates when the tightest bounds are found, that is, applying rules produce no new bounds, or when inconsistency is encountered, that is, a lower bound is larger than an upper bound.

For this example, two mappings (inf_1, sup_1) on $\mathcal{L}(\mathcal{A})$ are initialized from incidence function i with $inf_1(\phi) = sup_1(\phi) = i(\phi)$, $inf_1(\neg(\phi \wedge \neg\psi)) = sup_1(\neg(\phi \wedge \neg\psi)) = i(\neg(\phi \wedge \neg\psi))$.

In this case a constraint set is required to apply the inference rules and the constraint set is $\mathcal{L}(\mathcal{A} \cup \{\psi\}) = \{\phi, \psi, \neg\phi, \neg\psi, \phi \wedge \psi, \neg(\neg\phi \wedge \neg\psi), \neg\phi \wedge \psi, \phi \wedge \neg\psi, \neg(\neg\phi \wedge \psi), \neg(\phi \wedge \neg\psi), \neg(\phi \wedge \psi), \neg\phi \wedge \neg\psi\}$ as shown in Table 2.2.

Here $\neg(\phi \wedge \neg\psi) = \phi \rightarrow \psi$.

All the formulae used in Table 2.2 have been rewritten in the canonical form.

We will now see how to fire inference rules to update the bounds of those formulae.

Based on the incidence calculus inference procedure in Section 2.2.3, it is not difficult to see that there are no pairs of rules which will change both inf_1 and sup_1 of any formula in Table 2.2. So, changing one bound of a formula at a time is the only option left.

Scanning through formulae in Table 2.2, it is easy to see that the bounds of ϕ and $\neg\phi$ cannot be changed anymore. So we try to tighten the bounds of ψ first. Rules And_3 and And_4 are applied, because Not_1 and Not_2 have no impact on ψ which produce a better lower bound but keep the upper bound unchanged:

$$\begin{aligned}
sup_2(\psi) &= (sup_1(\phi \wedge \psi) \cup (\mathcal{W} \setminus inf_1(\phi))) \cap sup_1(\psi) && \text{by rule } \mathrm{And}_3 \\
&= (\{a\} \cup (\{a, b\} \setminus \{a\})) \cap \{a, b\} \\
&= \{a, b\} \\
inf_2(\psi) &= inf_1(\phi \wedge \psi) \cup inf_1(\psi) && \text{by rule } \mathrm{And}_4 \\
&= \{a\} \cup \emptyset \\
&= \{a\}
\end{aligned}$$

Table 2.2: Initial assignment of bounds on elements in the constraint set *Consecutive assignments are numbered as* 1, 2, *etc. The formulae in* $\mathcal{L}(\mathcal{A})$ *are assigned lower and upper bounds equal to their incidences. All non-axioms are assigned the default upper and lower bounds of* $\mathcal{W}$ *and* $\emptyset$*, respectively.*

Formula	inf_1	sup_1
ϕ	$\{a\}$	$\{a\}$
$\neg\phi$	$\{b\}$	$\{b\}$
ψ	$\emptyset$	$\{a,b\}$
$\neg\psi$	$\emptyset$	$\{a,b\}$
$\phi \wedge \psi$	$\{a\}$	$\{a\}$
$\neg\phi \wedge \psi$	$\emptyset$	$\{a,b\}$
$\phi \wedge \neg\psi$	$\emptyset$	$\emptyset$
$\neg\phi \wedge \neg\psi$	$\emptyset$	$\{a,b\}$
$\neg(\neg\phi \wedge \psi)$	$\emptyset$	$\{a,b\}$
$\neg(\neg\phi \wedge \neg\psi)$	$\emptyset$	$\{a,b\}$
$\neg(\phi \wedge \psi)$	{b}	$\{b\}$
$\neg(\phi \wedge \neg\psi)$	$\{a,b\}$	$\{a,b\}$

The second assignment, consisting of pair (inf_2, sup_2) is derived from the base assignment (inf_1, sup_1) through (And_3, And_4).

The next formula to be considered is $\neg\psi$ on which both pairs of rules (Not_3, Not_4) and $(And_3,\ And_4)$ are applicable and give the same results $inf_3(\neg\psi) = \emptyset$ and $sup_3(\neg\psi) = \{b\}$.

We only show the application of (And_3, And_4) here and leave the application of the first pair to the reader:

$$\begin{aligned}
sup_3(\neg\psi) &= (sup_2(\phi \wedge \neg\psi) \cup (\mathcal{W} \setminus inf_2(\phi))) \cap sup_2(\neg\psi) && \text{by rule And}_3 \\
&= (\emptyset \cup (\{a,b\} \setminus \{a\})) \cap \{a,b\} \\
&= \{b\} \\
inf_3(\neg\psi) &= inf_2(\phi \wedge \neg\psi) \cup inf_2(\neg\psi) && \text{by rule And}_4 \\
&= \emptyset \cup \emptyset \\
&= \emptyset
\end{aligned}$$

Table 2.3 with third assignment (inf_3, sup_3) shows the revised result of Table 2.2 by applying two pairs of inference rules.

Table 2.3 reflects the third assignment with mappings inf_3, and sup_3.

Similarly, pairs of rules are applied to all possible formulae for which the bounds can be further tightened.

The following list of tighter bounds are obtained after exhaustive applications of these rules (the final assignment is given in Table 2.4):

Table 2.3: Revised assignment of bounds on relevant formulae

The bounds of those formulae that have been changed from Table 2.2 have been labelled with the rule of inference that made the change and with the number of the base assignment.

Formula	inf_3	sup_3
ϕ	$\{a\}$	$\{a\}$
$\neg\phi$	$\{b\}$	$\{b\}$
ψ	$\{a\}(Add_4, 1)$	$\{a, b\}$
$\neg\psi$	$\emptyset$	$\{b\}(Add_3, 2)$
$\phi \wedge \psi$	$\{a\}$	$\{a\}$
$\neg\phi \wedge \psi$	$\emptyset$	$\{a, b\}$
$\phi \wedge \neg\psi$	$\emptyset$	$\emptyset$
$\neg\phi \wedge \neg\psi$	$\emptyset$	$\{a, b\}$
$\neg(\neg\phi \wedge \psi)$	$\emptyset$	$\{a, b\}$
$\neg(\neg\phi \wedge \neg\psi)$	$\emptyset$	$\{a, b\}$
$\neg(\phi \wedge \psi)$	{b}	$\{b\}$
$\neg(\phi \wedge \neg\psi)$	$\{a, b\}$	$\{a, b\}$

Table 2.4: Final assignment of bounds

Formula	inf_7	sup_7
ϕ	$\{a\}$	$\{a\}$
$\neg\phi$	$\{b\}$	$\{b\}$
ψ	$\{a\}(Add_4, 1)$	$\{a, b\}$
$\neg\psi$	$\emptyset$	$\{b\}(Add_3, 2)$
$\phi \wedge \psi$	$\{a\}$	$\{a\}$
$\neg\phi \wedge \psi$	$\emptyset$	$\{b\}(And_5, 3)$
$\phi \wedge \neg\psi$	$\emptyset$	$\emptyset$
$\neg\phi \wedge \neg\psi$	$\emptyset$	$\{b\}(And_5, 4)$
$\neg(\neg\phi \wedge \psi)$	$\{a\}(Not_3, 5)$	$\{a, b\}$
$\neg(\neg\phi \wedge \neg\psi)$	$\{a\}(Not_3, 6)$	$\{a, b\}$
$\neg(\phi \wedge \psi)$	{b}	$\{b\}$
$\neg(\phi \wedge \neg\psi)$	$\{a, b\}$	$\{a, b\}$

by rule And_5 :

$$\begin{aligned} sup_4(\neg\phi \wedge \psi) &= sup_3(\neg\phi) \cap sup_3(\psi) \cap sup_3(\neg\phi \wedge \psi) \\ &= (\{b\} \cap \{a,b\}) \cap \{a,b\} \\ &= \{b\} \end{aligned}$$

by rule And_6 :

$$\begin{aligned} inf_4(\neg\phi \wedge \psi) &= (inf_3(\neg\phi) \cap inf_3(\psi)) \cup inf_3(\neg\phi \wedge \psi) \\ &= (\{b\} \cap \{a\}) \cup \emptyset \\ &= \emptyset \end{aligned}$$

by rule And_5 :

$$\begin{aligned} sup_5(\neg\phi \wedge \neg\psi) &= sup_4(\neg\phi) \cap sup_4(\neg\psi) \cap sup_4(\neg\phi \wedge \neg\psi) \\ &= (\{b\} \cap \{b\}) \cap \{a,b\} \\ &= \{b\} \end{aligned}$$

by rule And_6 :

$$\begin{aligned} inf_5(\neg\phi \wedge \neg\psi) &= (inf_4(\neg\phi) \cap inf_4(\neg\psi)) \cup inf_4(\neg\phi \wedge \neg\psi) \\ &= (\{b\} \cap \emptyset) \cup \emptyset \\ &= \emptyset \end{aligned}$$

by rule Not_3 :

$$\begin{aligned} sup_6(\neg(\neg\phi \wedge \psi)) &= (\mathcal{W} \setminus inf_5(\neg\phi \wedge \psi)) \cap sup_5(\neg(\neg\phi \wedge \psi)) \\ &= (\{a,b\} \setminus \emptyset) \cap \{a,b\} \\ &= \{a,b\} \end{aligned}$$

by rule Not_4 :

$$\begin{aligned} inf_6(\neg(\neg\phi \wedge \psi)) &= (\mathcal{W} \setminus sup_5(\neg\phi \wedge \psi)) \cup inf_5(\neg(\neg\phi \wedge \psi)) \\ &= (\{a,b\} \setminus \{b\}) \cup \emptyset \\ &= \{a\} \end{aligned}$$

by rule Not_3 :

$$\begin{aligned} sup_7(\neg(\neg\phi \wedge \neg\psi)) &= (\mathcal{W} \setminus inf_6(\neg\phi \wedge \neg\psi)) \cap sup_6(\neg(\neg\phi \wedge \neg\psi)) \\ &= (\{a,b\} \setminus \emptyset) \cap \{a,b\} \\ &= \{a,b\} \end{aligned}$$

by rule Not_4 :

$$\begin{aligned} inf_7(\neg(\neg\phi \wedge \neg\psi)) &= (\mathcal{W} \setminus sup_6(\neg\phi \wedge \neg\psi)) \cup inf_6(\neg(\neg\phi \wedge \neg\psi)) \\ &= (\{a,b\} \setminus \{b\}) \cup \emptyset \\ &= \{a\} \end{aligned}$$

Therefore, the bounds of ψ lie between $\{a\}$ and $\{a, b\}$, which is the same result obtained using equations in Approach 1.

$\diamond$

Among all the formulae in Table 2.2, some formulae, such as $\neg(\neg\phi \wedge \neg\psi)$, can be rewritten as $\phi \vee \psi$.

If the disjunctive form of a formula is in use, different inference rules, such as Or_5 or Or_6, ought to be applied. No matter which form a formula is in and which rule is to apply, the final result remains the same. We leave the reader to reconstruct this example by replacing some formulae with their disjunctive forms and applying corresponding inference rules.

Example 2.3

Assume that an initial inf and sup are given for each formula in a subset $\mathcal{A} = \{\delta_1 \vee \delta_2, \delta_1 \vee \delta_3, \delta_1, \delta_3\}$ of $\mathcal{L}(P)$ as:

$$
\begin{array}{ll}
inf(\delta_1 \vee \delta_2) = \{w_1, w_4\}, & sup(\delta_1 \vee \delta_2) = \mathcal{W}; \\
inf(\delta_1) = \emptyset, & sup(\delta_1) = \{w_1, w_2, w_3\}; \\
inf(\delta_1 \vee \delta_3) = \{w_1, w_2\}, & sup(\delta_1 \vee \delta_3) = \mathcal{W}; \\
inf(\delta_3) = \emptyset, & sup(\delta_3) = \{w_3, w_5\},
\end{array}
$$

where $\mathcal{W} = \{w_1, w_2, \ldots, w_n\}$.

For simplicity, it is also assumed that δ_j are the basic elements in Definition 2.2. So, given any two distinct δ_j and δ_l, $\delta_j \wedge \delta_l = false$.

This example differs from Example 2.2 because inf and sup are defined from $\mathcal{A}$ to $2^{\mathcal{W}}$ without giving any incidence function. In this case the lower and upper bound equations, (2.1) and (2.2), cannot be used directly.

Assume that we want to know the bounds of incidences of formulae in

$$S = \{\delta_1, \delta_2, \delta_3, \delta_1 \vee \delta_2, \delta_1 \vee \delta_3, \delta_2 \vee \delta_3\},$$

inf and sup should be extended to be a pair of initial assignments (inf_1, sup_1) on S as shown in Table 2.5.

In this initial assignment, $\emptyset$ and $\mathcal{W}$ as lower and upper incidence bounds are added as default.

Starting from the initial assignment, we apply pairs of Or or Not inference rules on these formulae until no further tighter bounds are available (Table 2.6 contains the final assignment):

Table 2.5: The initial assignment of lower and upper bounds of incidences

Here, in order to simplify the statement, some formulae in the constraint set $\mathcal{L}(\mathcal{A} \cup S)$ which have no effect on the bounds of formulae in which we are interested have been ignored.

Formula	inf_1	sup_1
δ_1	$\emptyset$	$\{w_1, w_2, w_3\}$
δ_2	$\emptyset$	$\mathcal{W}$
δ_3	$\emptyset$	$\{w_3, w_5\}$
$\delta_1 \vee \delta_2$	$\{w_1, w_4\}$	$\mathcal{W}$
$\delta_1 \vee \delta_3$	$\{w_1, w_2\}$	$\mathcal{W}$
$\delta_2 \vee \delta_3$	$\emptyset$	$\mathcal{W}$

$$
\begin{aligned}
inf_2(\delta_1) &= (inf_1(\delta_1 \vee \delta_3) \cap (\mathcal{W} \setminus sup_1(\delta_3)) \cup inf_1(\delta_1) && \text{by rule Or}_1 \\
&= (\{w_1, w_2\} \cap (\mathcal{W} \setminus \{w_3, w_5\})) \cup \emptyset \\
&= \{w_1, w_2\} \\
sup_2(\delta_1) &= sup_1(\delta_1 \vee \delta_3) \cap sup_1(\delta_1) && \text{by rule Or}_2 \\
&= \mathcal{W} \cap \emptyset \\
&= \emptyset \\
inf_3(\delta_2) &= (inf_2(\delta_1 \vee \delta_2) \cap \mathcal{W} \setminus sup_2(\delta_1)) \cup inf_2(\delta_2) && \text{by rule Or}_3 \\
&= (\{w_1, w_4\} \cap (\mathcal{W} \setminus \{w_1, w_2, w_3\})) \cup \mathcal{W} \\
&= \{w_4\} \\
sup_3(\delta_2) &= sup_2(\delta_1 \vee \delta_2) \cap sup_2(\delta_2) && \text{by rule Or}_4 \\
&= \mathcal{W} \cap \mathcal{W} \\
&= \mathcal{W} \\
inf_4(\delta_1 \vee \delta_2) &= inf_3(\delta_1) \cup inf_3(\delta_2) \cup inf_3(\delta_1 \vee \delta_2) && \text{by rule Or}_5 \\
&= \{w_1, w_2\} \cup \{w_4\} \cup \{w_1, w_4\} \\
&= \{w_1, w_2, w_4\} \\
sup_4(\delta_1 \vee \delta_2) &= (sup_3(\delta_1) \cup sup_3(\delta_2)) \cap sup_3(\delta_1 \vee \delta_2) && \text{by rule Or}_6 \\
&= (\{w_1, w_2, w_3\} \cup \mathcal{W}) \cap \mathcal{W} \\
&= \mathcal{W} \\
inf_5(\delta_1 \vee \delta_3) &= inf_4(\delta_1) \cup inf_4(\delta_3) \cup inf_4(\delta_1 \vee \delta_3) && \text{by rule Or}_5 \\
&= \{w_1, w_2\} \cup \emptyset \cup \{w_1, w_2\} \\
&= \{w_1, w_2\} \\
sup_5(\delta_1 \vee \delta_3) &= (sup_4(\delta_1) \cup sup_4(\delta_3)) \cap sup_4(\delta_1 \vee \delta_3) && \text{by rule Or}_6 \\
&= (\{w_1, w_2, w_3\} \cup \{w_3, w_5\}) \cap \mathcal{W} \\
&= \{w_1, w_2, w_3, w_5\} \\
inf_6(\delta_2 \vee \delta_3) &= inf_5(\delta_2) \cup inf_5(\delta_3) \cup inf_5(\delta_2 \vee \delta_3) && \text{by rule Or}_5 \\
&= \{w_4\} \cup \emptyset \cup \emptyset \\
&= \{w_4\} \\
sup_6(\delta_2 \vee \delta_3) &= (sup_5(\delta_2) \cup sup_5(\delta_3)) \cap sup_5(\delta_2 \vee \delta_3) && \text{by rule Or}_6 \\
&= (\mathcal{W} \cup \{w_3, w_5\}) \cap \mathcal{W} \\
&= \mathcal{W} \\
sup_7(\delta_2) &= (\mathcal{W} \setminus inf_6(\neg\delta_2)) \cap sup_6(\delta_2) && \text{by rule Not}_1 \\
&= (\mathcal{W} \setminus \{w_1, w_2\}) \cap \mathcal{W} \\
&= \{w_3, w_4, w_5\} \\
inf_7(\delta_2 \vee \delta_3) &= (\mathcal{W} \setminus sup_6(\neg(\delta_2 \vee \delta_3))) \cup inf_6(\delta_2 \vee \delta_3)) && \text{by rule Not}_2 \\
&= (\mathcal{W} \setminus \{w_1, w_2, w_3\}) \cup \{w_4\} \\
&= \{w_4, w_5\} \\
sup_7(\delta_2 \vee \delta_3) &= (\mathcal{W} \setminus inf_6(\neg(\delta_2 \vee \delta_3))) \cap sup_6(\delta_2 \vee \delta_3) && \text{by rule Not}_1 \\
&= (\mathcal{W} \setminus \{w_1, w_2\}) \cap \mathcal{W} \\
&= \{w_3, w_4, w_5\}
\end{aligned}
$$

Table 2.6: Final assignment of lower and upper bounds of incidence

Formula	inf_7		$\sup_7$
δ_1	$\{w_1, w_2\}$	$(Or_1, 1)$	$\{w_1, w_2, w_3\}$
δ_2	$\{w_4\}$	$(Or_3, 2)$	$\{w_3, w_4, w_5\}(Not_1, 6)$
δ_3	$\emptyset$		$\{w_3, w_5\}$
$\delta_1 \vee \delta_2$	$\{w_1, w_2, w_4\}$	$(Or_5, 3)$	$\mathcal{W}$
$\delta_1 \vee \delta_3$	$\{w_1, w_2\}$		$\{w_1, w_2, w_3, w_5\}(Or_6, 4)$
$\delta_2 \vee \delta_3$	$\{w_4, w_5\}$	$(Not_2, 6)$	$\{w_3, w_4, w_5\}(Not_1, 6)$

This example is a revised version of an example in [Wong et al, 1992] in which the interval based method is used to reason with uncertain information, where the same result was given.

So far, an incidence calculus theory has been taken as the starting point, and the information it carries has been used to make inference. The inference of incidences, or bounds of incidences, is purely symbolic. This is done through the function i between a set of axioms $\mathcal{A}$ and a set of possible worlds.

The calculation of probability (or bounds) of a formula is simply done through the calculation of the probability of its incidence set (or its bounds of incidences). Obviously, obtaining the incidence set (or bounds) of a formula is the main step in this inference procedure, and this main step is in the form of a symbolic reasoning pattern.

2.4 Assigning Incidences to Formulae

In previous sections, we assume that an incidence calculus theory (or at least bounds of incidences) is known, with incidences assigned to axioms.

Probabilities of axioms are calculated through their incidences. However, it is very likely that probabilities are directly associated with a set of axioms without specifying the incidence functions, such as in pure numerical reasoning theories. In this case we need to recover the incidence assignment on axioms from the probability assignment. This procedure is called the *assigning of incidences to formulae.*

In [Bundy, 1992], it is assumed that the set of possible worlds is fixed, for instance 100 possible worlds, and the probability on each of them is equally distributed as, e.g., $1/100$ for every w. Under this assumption, the Monte Carlo method [Corlett and Todd, 1985] was used to divide these possible worlds into groups to satisfy the given probability distribution on the set of axioms. This method has been implemented in [McLean, 1992] and [McLean *et al.*, 1995].

Alternatively, a depth first approach on an incidence assignment tree has also been developed in [McLean *et al.*, 1995] to solve the same problem under the same assumptions.

Using these approaches, multiple consistent incidence assignments may be found given an initial probability distribution.

Here, we only briefly introduce these two approaches while for a detailed discussion and comparison of the two approaches we refer the reader to the work [McLean *et al.*, 1995].

A tree for incidence assignment

Both assignment methods studied in [McLean *et al.*, 1995] can be viewed as searching techniques for a tree. We will introduce this tree first.

Given a subset $\mathcal{A} = \{\phi_1, \phi_2, \ldots, \phi_n\}$ of a propositional language $\mathcal{L}(P)$, a probability space $(\mathcal{W}, 2^{\mathcal{W}}, \mu)$ and a mapping function $u : \mathcal{A} \to [0, 1]$, we construct a tree whose vertices are assignments, with the root associated with assignment (inf_0, sup_0) defined by:

$$\begin{aligned} inf_0(\psi) &= \emptyset, \\ sup_0(\psi) &= \mathcal{W} \end{aligned}$$

for all $\psi \in sf(\mathcal{A})$.

Let (inf, sup) be a vertex at level r, then for each $E \in 2^{\mathcal{W}}$ with

$$inf(\phi_r) \subseteq E \subseteq sup(\phi_r), u(\phi_r) = \mu(E),$$

there is a subtree of (inf, sup) whose root (inf_r, sup_r) is obtained by applying the incidence calculus inference procedure to the assignment (inf_q, sup_q) defined by

$$\begin{array}{ll} inf_q(\psi) = inf(\psi), & \forall \psi \in sf(\mathcal{A}) \setminus \{\phi_r\}, \\ sup_q(\psi) = sup(\psi), & \forall \psi \in sf(\mathcal{A}) \setminus \{\phi_r\}, \\ inf_q(\phi_r) = E, & \\ sup_q(\phi_r) = E, & \end{array}$$

with a queue of rules containing all inference rules having $inf_r(\phi_r)$ or $sup_r(\phi_r)$ in their contribution.

The incidence calculus inference procedure has an extra check after a rule is fired to see if there is any inconsistency, such as $inf_{qn}(\phi) \not\subseteq sup_{qn}(\phi)$. If there is, then terminate the procedure and claim a failure; otherwise, continue the process.

Depth first incidence assignment algorithm

A depth first search is performed on an incidence assignment tree (designed above) until either a consistent assignment is found or all leaves have been checked without a suitable assignment being discovered.

Example 2.4 (from [McLean *et al.*, 1995])

Let a set of possible worlds $\mathcal{W}$ be $\mathcal{W} = \{0,1,2,3,4,5,6,7,8,9\}$ with equal probabilities on all elements, i.e., $\mu(w) = 0.1$. Let $\mathcal{A} = \{x, y\}$ with $u(x) = 0.8$ and $u(y) = 0.2$.

Then there are 10!/(8!2!)=45 possible choices for $i(x)$ with $\mu(i(x)) = u(x)$. Similarly, there are 45 possible choices for $i(y)$ with $\mu(i(y)) = u(y)$. Thus, there are in total 2025 incidence functions compatible with the given uncertainty.

Some of the examples of consistent incidence assignments are:

$$i_1(x) = \{0,1,2,3,4,5,6,7\} \quad i_1(y) = \{8,9\},$$
$$i_2(x) = \{0,1,2,3,4,5,6,7\} \quad i_2(y) = \{7,8\},$$
$$i_3(x) = \{0,1,2,3,4,5,6,7\} \quad i_3(y) = \{6,7\}.$$

These three incidence assignments give three different estimates for the uncertainty of $x \wedge y$:

$$\mu(i_1(x \wedge y)) = 0.0,$$
$$\mu(i_2(x \wedge y)) = 0.1,$$
$$\mu(i_3(x \wedge y)) = 0.2.$$

These three estimates for the uncertainty of $x \wedge y$ divide all possible incidence assignments into three groups. It is calculated that there are 45 incidence assignments in the first group with the uncertainty of $x \wedge y$ as 0.0; 720 in the second group with uncertainty 0.1; and 1260 in the third group with uncertainty 0.2.

◊

A number of shortcomings associated with this method were also revealed [McLean *et al.*, 1995]:

- It is possible to have two leafs (inf_1, sup_1) and (inf_2, sup_2) of the assignment tree, each of which is compatible with the initial numerical assignment but for which the two intervals:

$$[\mu(inf_1(\phi)), \mu(sup_1(\phi))]$$

$$[\mu(inf_2(\phi)), \mu(sup_2(\phi))]$$

are disjoint.

- This method may be very inefficient since time may be wasted on discovering a large number of assignments which are permutations of a few basic ones.
- A search for a single consistent leaf may terminate with an unrepresentative sample which will then lead to poor estimates of uncertainties on other formulae.

Monte Carlo incidence assignment method

McLean ([McLean, 1992]) implemented a Monte Carlo based approach that has initially introduced in [Corlett and Todd, 1985] and further developed in [Bundy, 1992]. This approach uses a suitable random choice of subtree at each vertex to carry out the search of the incidence assignment tree. This assignment method, as McLean discussed, suffers from the same three limitations as the depth first incidence assignment algorithm (given above).

Example 2.5 (from [McLean, 1992] which is a re-structured form of that from ([Kyburg, 1987]))

Suppose that the initial numerical assignment $\mathcal{A} = \{a \wedge b \wedge c, a \wedge c, b \wedge c, a, b, c\}$ (in this order) is:

$$\begin{aligned}
&\mu(a \wedge b \wedge c) = 0.2,\\
&\mu(a \wedge c) = 0.4,\\
&\mu(b \wedge c) = 0.2,\\
&\mu(a) = 0.8,\\
&\mu(b) = 0.6,\\
&\mu(c) = 0.5,
\end{aligned}$$

with a fixed set of possible worlds $W = \{0, 1, 2, 3, 4, 5, 6, 7, 8, 9\}$ which has $\mu(w) = 1/10$ for any $w \in \mathcal{W}$.

Applying the depth first incidence assignment algorithm to the axioms in set $\mathcal{A}$ (in that order), it takes 10.7 (seconds) to find the first consistent assignment at case 6 (the sixth try). However, if the axioms are reordered as $\{a, b, c, a \wedge c, b \wedge c, a \wedge b \wedge c\}$, then it tries at least 52 cases taking 70.6 (seconds) before a consistent assignment is found.

Applying the Monte Carlo incidence assignment method to this example, a consistent assignment can be found when the axioms are ordered as $\{a \wedge b \wedge$

$c, a \wedge c, b \wedge c, a, b, c\}$ (with no precise time given). When the order is changed to $\{a, b, c, a \wedge c, b \wedge c, a \wedge b \wedge c\}$, then no consistent assignment can be found at all, no matter how long it takes.

Example 2.6

A more complicated example used by McLean [McLean, 1992] is as follows.

Assume that a numerical assignment, which assigns a probability to an axiom to 2 places of decimals, on a set of axioms $\mathcal{A} = \{a \wedge b \wedge c, a \wedge c, b \wedge c, a, b, c\}$ (in this order) is:

$$\begin{aligned}
&\mu(a \wedge b \wedge c) = 0.18,\\
&\mu(a \wedge b) = 0.52,\\
&\mu(a \wedge c) = 0.35,\\
&\mu(b \wedge c) = 0.22,\\
&\mu(a) = 0.760,\\
&\mu(b) = 0.640,\\
&\mu(c) = 0.480,
\end{aligned}$$

with the assumption that W has 100 elements and for every $w \in \mathcal{W}$, $\mu(w) = 1/100$.

Using the depth first incidence assignment algorithm, it takes 374.52 (seconds) to find the first consistent assignment and it takes 355.06 (seconds) using the Monte Carlo incidence assignment method. The experimental results show that both the algorithms are rather slow.

There is only one extra axiom $(a \wedge b)$ in Example 2.6 than in Example 2.5. However, these two examples present rather different sets of execution time. In Example 2.5, 10 possible worlds would be enough to produce the given probabilistic assignment while Example 2.6 requires at least 100 possible worlds, in order to produce a probability to 2 places of decimals. Because the size of the set of possible worlds is larger in Example 2.6, there are more possible ways of dividing these possible worlds into subsets to satisfy the given probability distribution. That is why the two apparently similar examples have enormous difference in their execution time.

2.5 Summary

Incidence calculus is a logic for probabilistic reasoning intended to overcome the deficiency in pure numerical uncertainty approaches. Its distinctive feature over other numerical methods is the indirect encoding of probabilities to formulae via incidence sets. This feature allows incidence calculus to be truth functional, that is, the incidences of a compound formula can be calculated directly from the incidences of its constituent components.

When only bounds of incidences, rather than incidence functions, are known on a set of axioms, there is a procedure allowing bounds on non-axioms to be derived using a number of inference rules. This procedure is called the Legal Assignment Finder. The basis of applying the inference rules in the Legal Assignment Finder is the nature of truth functionality of incidence calculus.

The method of using bounds (or intervals), either symbolic (inf, sup) or numerical $(Prob_*, Prob^*)$, to describe uncertain information has been adopted in several mechanisms for reasoning under uncertainty.

In [Wong et al, 1992], a unified interval structure is proposed which is proved to be compatible with the lower and upper approximation of a set in *rough sets theory* [Pawlak, 1991], with the lower and upper bounds of incidences in incidence calculus and with belief and plausibility functions in DS theory. These theories which all obey the principles in the interval structure may have some inter-connections with each other.

The relationship between incidence calculus and DS theory will be studied in Chapter 7 and that of incidence calculus and rough sets will be investigated in Chapter 10.

Apart from its merit in unifying symbolic and numerical reasoning into one structure, however, incidence calculus suffers from a few limitations, due to the requirement of incidence function i being truth functional.

For instance, consider function i' defined as:

$$i'(rainy) = \{mon, tues\}, \qquad i'(windy) = \{wed\},$$

and

$$i'(rainy \vee windy) = \{mon, tues, wed, thur\}.$$

This function, i', is *not* an incidence function as $i'(rainy \vee windy) \neq i'(rainy) \cup i'(windy)$. The evidence only suggests that on *thur* it will be *rainy* or *windy*. It gives no further specification as to whether it will be *rainy*, *windy* or both. So we cannot put *thur* in either $i'(rainy)$ or $i'(windy)$. There is no corresponding incidence calculus theory to represent this piece of information.

In summary, there are three major drawbacks in the original incidence calculus:

1. Limition in representing ignorance.

Property $i(\neg\phi) = \mathcal{W} \setminus i(\phi)$ of i requires that the elements in $\mathcal{W}$ must be distributed into either $i(\phi)$ or $i(\neg\phi)$.

If both $i(\phi)$ and $i(\neg\phi)$ are specified, respectively, then $i(\phi) \cup i(\neg\phi)$ should be the whole set $\mathcal{W}$.

Property $i(\phi_1 \vee \phi_2) = i(\phi_1) \cup i(\phi_2)$ says that if $i(\phi_1)$, $i(\phi_2)$ and $i(\phi_1 \vee \phi_2)$ are all known, the possible worlds in $i(\phi_1 \vee \phi_2)$ can be split into two groups (not necessarily disjoint) $i(\phi_1)$ and $i(\phi_2)$. $i(\phi_1 \vee \phi_2)$ carries no more information than the union of $i(\phi_1)$ and $i(\phi_2)$.

When a real situation fails to meet either of the above two properties, that is, a real situation implies ignorance, incidence calculus theories are unable to describe it.

2. Absence of an efficient algorithm for constructing incidence functions from numerical assignments.

   The indirect encoding feature in the original incidence calculus requires the explicit definition of an incidence function i.

   In many applications, however, numerical values are directly assigned to formulae without specifying the corresponding incidence function i. To apply incidence calculus in these circumstances, it is necessary to construct a proper incidence function i from the numerical assignments.

   The two assignment methods reported in [McLean *et al.*, 1995] are not sufficient for this purpose due to their inefficiency. Furthermore, it is not necessary to discover all the possible consistent assignments because these assignments can well be represented by only a few basic assignments.

   Therefore, an efficient algorithm for constructing incidence functions from numerical assignments is crucial in applying incidence calculus. This limitation has also been pointed out in [Frisch and Haddawy, 1994].

3. Failing to deal with multiple pieces of evidence.

   Combining multiple pieces of evidence is vital in every mechanism for reasoning under uncertainty. A successful reasoning approach should provide a convenient and powerful mechanism to deal with accumulated information, like conditioning or updating methods in various probability related approaches, or Dempster's combination rule in DS theory.

In the original incidence calculus, it is not clear how to cope with multiple sources of information. The inference mechanism is designed to deal with one incidence calculus theory only. The basic inference mechanism in the theory is to ultimately infer the incidence bounds for a formula. It lacks the ability to combine the impact of several sources of information.

Chapter 3

Generalizing Incidence Calculus

At the end of Chapter 2, we discussed three major limitations of the original incidence calculus. These drawbacks will be dealt with individually in three consecutive Chapters 3, 4 and 5.

The first drawback is its limited ability in representing ignorance. Ignorance is usually caused by lacking knowledge or evidence that affects results of inference or decision making.

The requirement of being truth functional is the cause of this limitation. This requirement excludes incidence calculus from being applied in some situations when ignorance exists. In this chapter, we aim at improving incidence calculus so that it is possible to represent ignorance.

3.1 Generalized Incidence Calculus Theories

In order for incidence calculus to be able to represent a situation involving ignorance, we first need to extend two-valued logic to three-valued logic.

In the original incidence calculus, when an incidence function i is defined, the underlying propositional logic is two-valued. Any atomic proposition, and hence formula, will have either $true$ or $false$ (not both) as its truth value in a particular state of the world. This implies that if formula ϕ is not identified having truth value $true$ under a possible world w, then ϕ is recognized having truth value $false$ and $\neg\phi$ having truth value $true$.

Two-valued logic may not be suitable for situations where propositions describing causal future events are involved, i.e., *future contingents* (for details, see [Malinowski, 1993]). As a result, three-valued (or many-valued, multiple-valued) logic was proposed (e.g. [Malinowski, 1993]) in which an atomic proposition (or formula) has either $true$, $false$, or $unknown$ (or $undetermined$, or 1/2 as used in other books/articles) as its truth values.

Table 3.1: Truth table of five logical connectives in three-valued logic

q_1	q_2	$\neg q_1$	$q_1 \wedge q_2$	$q_1 \vee q_2$	$q_1 \rightarrow q_2$	$q_1 \leftrightarrow q_2$
true	*true*	*false*	*true*	*true*	*true*	*true*
true	*false*	*false*	*false*	*true*	*false*	*false*
true	1/2	*false*	1/2	*true*	1/2	1/2
false	*true*	*true*	*false*	*true*	*true*	*false*
false	*false*	*true*	*false*	*false*	*true*	*true*
false	1/2	*true*	*false*	1/2	*true*	1/2
1/2	*true*	1/2	1/2	*true*	*true*	1/2
1/2	*false*	1/2	*false*	1/2	1/2	1/2
1/2	1/2	1/2	1/2	1/2	*true*	*true*

Therefore, some important laws of two-valued logic are not tautologies of three-valued logic. For instance, $\phi \vee \neg\phi$ may not be a tautology in all possible states of the world, if there is a state in which both ϕ and $\neg\phi$ have truth value *unknown*.

The truth table for five logical connectives in two-valued logic is revised to accommodate the third value as shown in Table 3.1.

Based on three-valued logic, we modify Definition 2.4 as follows.

Definition 3.1: *Generalized Incidence Calculus Theory* (GICT)

A generalized incidence calculus theory (GICT) is a quintuple

$$< \mathcal{W}, \mu, P, \mathcal{A}, i >,$$

where:

- $\mathcal{W}$ *is a finite set of possible worlds.*
- *For all* $w \in \mathcal{W}$, $\mu(w)$ *is the probability of* w *and* $\mu(\mathcal{W}) = 1$, *where* $\mu(I) = \Sigma_{w \in I}\mu(w)$.
- P *is a finite set of atomic propositions.* $\mathcal{A}t$ *is the basic element set of* P. $\mathcal{L}(P)$ *is the language set of* P. *For any atomic proposition* $q \in P$, q *may have either true, false, or unknown (only one of them) as its truth value.*
- $\mathcal{A}$ *is a distinguished set of formulae in* $\mathcal{L}(P)$ *called the axioms of the theory. True and false are always taken as two axioms in* $\mathcal{A}$.
- i *is a function,* $i : \mathcal{A} \rightarrow 2^{\mathcal{W}}$. $i(\phi)$ *is to be thought of as the set of possible worlds in* $\mathcal{W}$ *in which* ϕ *has truth value true, i.e.,* $i(\phi) = \{w \in \mathcal{W} | \ w \models \phi\}$.

  Besides, $i(true) = \mathcal{W}$ *and* $i(false) = \emptyset$. $i(\phi)$ *is called the incidence set of* ϕ.

i defined in this way cannot be extended as a function from $\mathcal{L}(\mathcal{A})$ to $2^{\mathcal{W}}$, as some truth functionalities in two-valued logic are not valid anymore.

According to this revised definition, for $w \in \mathcal{W}$, if $w \notin i(\phi)$, it should not be concluded that $w \in i(\neg\phi)$. That is, it should not be assumed that $\neg\phi$ has truth value *true* when ϕ has truth value *false* in w. Therefore, equation $i(\phi) = \mathcal{W} \setminus i(\neg\phi)$ does not hold for i in a GICT.

Moreover, if $i(\phi), i(\psi)$ and $i(\phi \vee \psi)$ are all given, it is still possible that $i(\phi) \cup i(\psi) \subset i(\phi \vee \psi)$, because a particular possible world w may be known to support the joint statement $\phi \vee \psi$ but not each individual sub-statement (as in Example 3.1 given below).

Example 3.1

Suppose that there are ten delegates attending a meeting. The meeting will be held on a day next week for which all the delegates are asked to give their preferences. The meeting will be held on the day which is preferred by most of the delegates. Suppose that delegates 1 to 4 prefer only Monday, delegate 5 prefers Monday or Tuesday but not both days, the rest prefer only Tuesday.

Then a mapping function i in a GICT can be defined as

$$i(q_1) = \{a_1, a_2, a_3, a_4\},$$

$$i(q_2) = \{a_6, a_7, a_8, a_9, a_{10}\},$$

and

$$i(q_1 \vee q_2) = \{a_1, \ldots, a_5, \ldots, a_{10}\},$$

where q_1 stands for 'The meeting is held on Monday', q_2 for 'The meeting is on Tuesday'.

The corresponding GICT is

$$< W, \mu, P, \mathcal{A}, i >$$

where $\mathcal{W} = \{a_1, a_2, \ldots, a_{10}\}$ and $\mathcal{A} = \{q_1, q_2, q_1 \vee q_2, true, false\}$. a_l (for $j+1, \ldots, 10$) is a possible world in which delegate l holds his/her preference.

As $i(q_1 \wedge q_2) = \emptyset$, it is not necessary to include $q_1 \wedge q_2$ in $\mathcal{A}$ as an axiom.

Obviously, we have $i(q_1) \cup i(q_2) \subset i(q_1 \vee q_2)$ because a_5 cannot be put into either $i(q_1)$ or $i(q_2)$ alone.

If we put a_5 in $i(q_1)$ but not in $i(q_2)$, it does not precisely represent the meaning that a_5 could also attend the meeting on Tuesday if necessary. Similarly we cannot put a_5 in $i(q_2)$ but not in $i(q_1)$.

If a_5 is included in both $i(q_1)$ and $i(q_2)$, it is implied that a_5 is available on both days which conflicts with a_5's initial claim. So a_5 can only be included in $i(q_1 \vee q_2)$.

Proposition 3.1:

Let ϕ, ψ and $\phi \wedge \psi$ be three axioms in $\mathcal{A}$ in a GICT, then the following equation holds

$$i(\phi \wedge \psi) = i(\phi) \cap i(\psi). \tag{3.1}$$

Proof.

For all $w \in \mathcal{W}$, we have:

$w \in i(\phi \wedge \psi)$
$\Leftrightarrow w \models \phi \wedge \psi$ ($\phi \wedge \psi$ is true in w)
$\Leftrightarrow w \models \phi, w \models \psi$ (ϕ and ψ have truth values *true* in w)
$\Leftrightarrow w \in i(\phi), w \in i(\psi)$
$\Leftrightarrow w \in i(\phi \cap i(\psi)$

The non-logical notation $\Leftrightarrow$ is used to link the two logical expressions which are equivalent.

Proposition 3.1 suggests that although in a GICT i cannot be extended to a mapping between $\mathcal{L}(\mathcal{A})$ to $2^{\mathcal{W}}$, it can however be extended to a mapping between $\wedge(\mathcal{A})$ to $2^{\mathcal{W}}$, where $\wedge(\mathcal{A})$ denotes the language set which contains $\mathcal{A}$ and all the possible conjunctions of its elements.

For $\wedge_j \phi_j \in \wedge(\mathcal{A}) \setminus \mathcal{A}$ ($\phi_j \in \mathcal{A}$), we define $i(\wedge_j \phi_j) = \cap_j i(\phi_j)$. Therefore, the domain of i, the set of axioms $\mathcal{A}$, can always be extended to a set which is closed under the operator $\wedge$.

In the following, we always assume that the set of axioms $\mathcal{A}$ in a GICT has already been extended in this way. After a set $\mathcal{A}$ is extended to be closed under $\wedge$, theoretically, every formula $\wedge_j \phi_j$ should be in $\mathcal{A}$, when $\phi_j \in \mathcal{A}$.

If $i(\wedge_j \phi_j) = \emptyset$, nevertheless, it does not matter whether $\wedge_j \phi_j$ is in $\mathcal{A}$ as this formula has no effect on further inferences. However if $\wedge_j \phi_j$ is false, $i(\wedge_j \phi_j) = \cap_j i(\phi_j)$ must be empty; otherwise the information for constructing the function i is contradictory.

Under the principle of three-valued logic, incidence function i in a GICT has a number of properties as stated in the following propositions.

Proposition 3.2:

Let ϕ be an axiom in $\mathcal{A}$ in a GICT. If both $i(\phi)$ and $i(\neg\phi)$ are given, then

$$i(\phi) \cap i(\neg\phi) = \emptyset,$$

and

$$i(\phi) \cup i(\neg\phi) \subseteq \mathcal{W}.$$

Proof.

Assume $i(\phi) \cap i(\neg\phi) = W_1 \neq \emptyset$, then:

$$\begin{aligned}&\forall w \in W_1, w \in i(\phi) \cap i(\neg\phi)\\&\Rightarrow w \in i(\phi), w \in i(\neg\phi)\\&\Rightarrow \text{both } \phi \text{ and } \neg\phi \text{ have truth value } \mathit{true} \text{ in } w\\&\Rightarrow \text{Conflict.}\end{aligned}$$

So, $i(\phi) \cap i(\neg\phi) = \emptyset$ holds.
As $i(\phi) \subseteq \mathcal{W}$ and $i(\neg\phi) \subseteq \mathcal{W}$, it is obvious that $i(\phi) \cup i(\neg\phi) \subseteq \mathcal{W}$.

QED

Property $i(\phi) \cap i(\neg\phi) = \emptyset$ says that there are no possible worlds in which both formula ϕ and its negation $\neg\phi$ have truth value *true* at the same time.

Property $i(\phi) \cup i(\neg\phi) \subseteq \mathcal{W}$ indicates that there may be some possible worlds in which both formula ϕ and its negation $\neg\phi$ have truth value *unknown*. So these possible worlds will not be $i(\phi) \cup i(\neg\phi)$.

Proposition 3.3:

Let ϕ and ψ be two axioms in $\mathcal{A}$ in a GICT. If $i(\phi)$, $i(\psi)$, and $i(\phi \vee \psi)$ are known, then

$$i(\phi) \cup i(\psi) \subseteq i(\phi \vee \psi).$$

Proof.

Assume that $i(\phi) = W_1$, we have:

$$\begin{aligned}&\forall w \in W_1, w \in i(\phi)\\&\Rightarrow \phi \text{ is has truth value } \mathit{true} \text{ in possible world } w\\&\Rightarrow \phi \vee \psi \text{ has truth value } \mathit{true} \text{ in } w\\&\Rightarrow w \in i(\phi \vee \psi)\\&\Rightarrow i(\phi) \subseteq i(\phi \vee \psi).\end{aligned}$$

Similarly, we have $i(\psi) \subseteq i(\phi \vee \psi)$. Therefore, $i(\phi) \cup i(\psi) \subseteq i(\phi \vee \psi)$.

QED

This proposition tells us that given two formulae ϕ and ψ, it is very likely that the incidence set $i(\phi \vee \psi)$ contains more information than the union of $i(\phi)$ and $i(\psi)$.

In other word, there may exist a possible world w in which formula $\phi \vee \psi$ has truth value *true*, but formulae ϕ and ψ have truth value *unknown* for the time being, as shown in Example 3.1.

Proposition 3.4:

Let ϕ and ψ be two axioms in $\mathcal{A}$ in a GICT. If $\phi \models \psi$, then $i(\phi) \subseteq i(\psi)$.

Proof.

If $\phi \models \psi$, then we have:

$$\begin{aligned}
&\forall w \in i(\phi) \\
&\Rightarrow \phi \text{ has truth value } \textit{true} \text{ in possible world } w \\
&\Rightarrow \psi \text{ also has truth value } \textit{true} \text{ in } w \\
&\Rightarrow w \in i(\psi) \\
&\Rightarrow i(\phi) \subseteq i(\psi).
\end{aligned}$$

QED

This proposition states that if whenever ϕ is true and ψ is true, under any interpretation, then the incidence set of ϕ is just a subset of the incidence set of ψ.

For a formula $\phi \in \mathcal{L}(P) \setminus \mathcal{A}$, in general we can only define the lower and upper bounds of its incidence set using the functions i^* and i_* respectively in Definition 3.2.

Definition 3.2: *Lower and Upper Bounds of Incidences*

Given a set of axioms $\mathcal{A}$ in a GICT, for a formula $\phi \in \mathcal{L}(P) \setminus \mathcal{A}$, the lower and upper bounds of the incidence set of ϕ are defined as:

$$i_*(\phi) = \bigcup_{\psi \in \mathcal{A}, \psi \models \phi} i(\psi); \tag{3.2}$$

and

$$i^*(\phi) = \mathcal{W} \setminus i_*(\neg\phi). \tag{3.3}$$

We need to bear in mind that $\mathcal{A}$ is closed under $\wedge$.

The lower bound definition in equation (3.2) is the same as equation (2.1) in the original incidence calculus. However, equation (3.3) for the upper bound above uses the result of Proposition 2.2 instead of equation (2.2).

If we use equation (2.2) for the upper bound in GICTs, we cannot prove $i^*(\phi) = \mathcal{W} \setminus i_*(\neg\phi)$ after dropping some conditions on i in the original incidence calculus (check the proof again if you are not convinced!). On the other hand, $i^*(\phi) = \mathcal{W} \setminus i_*(\neg\phi)$ is compatible with other definitions of upper bounds in other theories, so the upper bound is defined in this way here.

The lower bound represents the set of possible worlds in which ϕ definitely has truth value *true* and the upper bound represents the set of possible worlds in which ϕ may have truth value *true*, i.e., the upper bound excludes those possible worlds in which ϕ definitely has *false* as its truth value.

Function $Prob_*(\phi) = \mu(i_*(\phi))$ gives the degree of our belief in ϕ and function $Prob^*(\phi) = \mu(i^*(\phi))$ represents the degree to which we fail to believe in $\neg\phi$. For a formula ϕ in $\mathcal{A}$, if $Prob_*(\phi) = Prob^*(\phi)$, then $Prob(\phi)$ is defined as $Prob_*(\phi)$ and called the probability of this formula.

$Prob_*(\phi)$ and $Prob^*(\phi)$ correspond to the lower and upper bounds of probabilities in probability theory. They are also equivalent to the belief and plausibility functions in DS theory [Liu and Bundy, 1994] (for more detail, see also Chapter 7).

In a situation where *true* and *false* are the only axioms in a GICT, for any formula $\phi \in \mathcal{L}(P) \setminus \mathcal{A}$, there is $i_*(\phi) = \emptyset$ and $i^*(\phi) = \mathcal{W}$. We know nothing about the whole situation. This gives the following definitions.

Definition 3.3: *The Representation of Total Ignorance*

Given a GICT $< \mathcal{W}, \mu, P, \mathcal{A}, i >$, *if* $\mathcal{A} = \{true, false\}$, *we say that this GICT represents total ignorance.*

Definition 3.4: *Conditional Probabilities*

Given a set of axioms $\mathcal{A}$ *in a GICT, if for every* $\phi \in \mathcal{A}$, $Prob_*(\phi) = Prob^*(\phi)$, *then we can define the conditional probability of* ϕ *given* ψ $(\psi \in \mathcal{A})$ *as:*

$$Prob(\phi \mid \psi) = \frac{Prob(\phi \wedge \psi)}{Prob(\psi)}. \tag{3.4}$$

When $Prob_*(\phi)$ and $Prob^*(\phi)$ are not equal, it is not possible to define $Prob$ on $\mathcal{A}$, or to define the conditional probability. We can only talk about the bounds of probabilities.

Therefore, given a formula $\phi \in \mathcal{L}(P)$, it is more natural to describe the lower bound of its probability rather than its probability. In the following, when we mention a numerical assignment on a language set, we always mean the lower bound of an unspecified probability measure.

Example 3.3 (Originally from [Fagin and Halpern, 1989a] and used in [Correa da Silva and Bundy, 1990]). We have the following situation:

> A person has four coats: two are blue with single-breasted, one is grey and double breasted and one is grey and single breasted. To choose which colour of coat to wear, this person tosses a (fair) coin. Once the colour is chosen, which specific coat is worn is determined by a mysterious procedure. What is the probability of the person wearing a single-breasted coat?

To solve this problem in generalized incidence calculus, we need to consider how to construct a GICT.

Let a set of propositions P be $P = \{grey, double\}$ where $grey$ stands for 'The coat is grey' and $double$ stands for 'The coat is double-breasted' and let a set of possible worlds be $\mathcal{W} = \{w_1, w_2\}$ in which blue and grey coats are worn respectively.

Then, we have

$$\mathcal{A}t = \{grey \wedge double, \neg grey \wedge double, grey \wedge \neg double, \neg grey \wedge \neg double\}.$$

Among these basic elements, $\neg grey \wedge double$ is always invalid, as there are no such coats among the four choices. Possible world w_1 supports formula $\neg grey \wedge \neg double$ and w_2 supports formula $(grey \wedge \neg double) \vee (grey \wedge double)$.

Therefore, we get a GICT, $< \mathcal{W}, \mu, P, \mathcal{A}, i >$, where:

$$\begin{aligned}
\mu(w_1) &= \mu(w_2) = 0.5, \\
\mathcal{A}Z &= \{\neg grey \wedge \neg double, \\
&\quad (grey \wedge \neg double) \vee \\
&\quad \vee (grey \wedge double), true, false\}, \\
i(\neg grey \wedge \neg double) &= \{w_1\}, \\
i((grey \wedge \neg double) \vee (grey \wedge double)) &= \{w_2\}, \\
i(true) &= \{w_1, w_2\}, \\
i(false) &= \emptyset.
\end{aligned}$$

So:

$$\begin{aligned}
i_*(\neg double) &= i(\neg grey \wedge \neg double), \\
i^*(\neg double) &= \mathcal{W} \setminus i_*(double) = \mathcal{W},
\end{aligned}$$

and

$$\begin{aligned}
Prob_*(\neg double) &= 0.5, \\
Prob^*(\neg double) &= 1.
\end{aligned}$$

The answer to the question is that the probability of the person wearing a single-breasted coat lies between 0.5 and 1.

$\diamond$

The main concern of this section has to generalize the original incidence calculus. We achieved this by using three-vaued propositional logic to replace two-valued logic in the original incidence calculus. As we have seen, GICTs are able to represent those situations where the original incidence calculus are powerless.

In the next section, we will see whether an incidence function is the only mapping function between a set of axioms and a set of possible worlds.

3.2 Basic Incidence Assignment

For each axiom $\phi \in \mathcal{A}$ of a GICT $< \mathcal{W}, \mu, P, \mathcal{A}, i >$, the incidence set $i(\phi)$ contains every possible world which is known to make ϕ true.

When $w \in i(\psi)$ and $\psi \models \phi$ ($\psi \neq \phi$), then $w \in i(\phi)$ (Proposition 3.4). Assume that $\psi_1, \ldots, \psi_j \in \mathcal{A}$ are all the axioms satisfying the condition $\psi_j \models \phi$ ($\psi_j \neq \phi$), then $i(\phi) \setminus \cup_j i(\psi_j)$ consists of those possible worlds which make only ϕ true without making any of ψ_j true when it is not empty.

We are particularly interested in ϕ when $i(\phi) \setminus \cup_j i(\psi_j)$ is not empty. What properties do these possible worlds possess?

Definition 3.5: *Function ii From i on* $\mathcal{A}$

Let i be the incidence assignment in a GICT, $< \mathcal{W}, \mu, P, \mathcal{A}, i >$, *then a new function* $ii : \mathcal{A} \to 2^{\mathcal{W}}$ *can be defined through the following equation:*

$$ii(\phi) = i(\phi) \setminus \cup_{\psi_j \models \phi, \psi_j \neq \phi} i(\psi_j),$$

for $\phi, \psi_j \in \mathcal{A}$.

It is possible that for some ψ, $ii(\psi) = \emptyset$. As we are more interested in those axioms with $ii(\psi) \neq \emptyset$, we have the following definition.

Definition 3.6 *Regulated ii*

Let ii be a function $ii : \mathcal{A} \to 2^{\mathcal{W}}$ *obtained from Definition 3.5.*

ii is said to have been regulated on domain $\mathcal{A}_0$ ($\mathcal{A}_0 \subseteq \mathcal{A}$) *iff:*

1. $false \in \mathcal{A}_0$, $ii(false) = \emptyset$,
2. $\forall \phi \in \mathcal{A}_0$ ($\phi \neq false$), $ii(\phi) \neq \emptyset$,
3. $\forall \psi \in \mathcal{A} \setminus \mathcal{A}_0$, $ii(\psi) = \emptyset$.

A regulated ii eliminates those axioms (except $false$) which have empty ii sets.

In practice, how do we extract function ii from a set of axioms? Algorithm A shown below provides a technical procedure for doing so while Proposition 3.5 and Theorem 1 together prove theoretically that the output of Algorithm A produces a unique function satisfying the condition in Definition 3.6.

Algorithm A: Extraction of function ii

input: *a GICT* $< \mathcal{W}, \mu, P, \mathcal{A}, i >$
output: *a function* $ii : \mathcal{A}_0 \rightarrow 2^{\mathcal{W}}$ *where* $\mathcal{A}_0 \subseteq \mathcal{A}$

Step 0: choose *a subset* $\mathcal{A}_0$ *of* $\mathcal{A} \setminus \{false\}$, *as* $\mathcal{A}_0 = \{\psi_1, \ldots, \psi_n\}$ *where* $\mathcal{A}_0$ *satisfies the condition*

$$\forall \psi_i \in \mathcal{A}_0, \forall \phi \in \mathcal{A}, \quad \phi \not\models \psi_i \ if \ \phi \neq \psi_i. \tag{3.5}$$

{Therefore, $\mathcal{A}_0$ *contains the "smallest"*[1] *formulae in* $\mathcal{A}$ *and* $\mathcal{A}_0$ *is not empty*[2].

In fact, we can find at least one element belonging to $\mathcal{A}_0$ *using the following procedure. For a formula* $\phi \in \mathcal{A} \setminus \{false\}$, *if* $\exists \psi_i \in \mathcal{A}, \psi_i \neq \phi$ *and* $\psi_i \models \phi$, *then we use* ψ_i *to replace* ϕ *and repeat the same procedure*[3] *until we obtain a formula* ψ_j *and we cannot find any formula which makes* ψ_j *true, then* ψ_j *will be in* $\mathcal{A}_0$. *}*

This procedure is:

Step 1: for each *formula* $\psi \in \mathcal{A}_0$ **do** $ii(\psi) \leftarrow i(\psi)$

Step 2: $\mathcal{A}' \leftarrow \mathcal{A} \setminus (\mathcal{A}_0 \cup \{false\})$

Step 3: if $\mathcal{A}'$ *is left with only tautologies* **then** *go to* **Step 6**.

Step 4: choose *a formula* ϕ_l *in* $\mathcal{A}'$ **such that for each** $\phi_j \in \mathcal{A}'$ **if** $\phi_j \neq \phi_l$, **then** $\phi_j \not\models \phi_l$ **end for**

$ii(\phi_l) \leftarrow i(\phi_l) \setminus \bigcup_{\psi_{lj} \in \mathcal{A}_0, \psi_{lj} \models \phi_l} ii(\psi_{lj})$

Step 5: $\mathcal{A}' \leftarrow \mathcal{A}' \setminus \{\phi_l\}$

if $ii(\phi_l) \neq \emptyset$ **then** $\mathcal{A}_0 \leftarrow \mathcal{A}_0 \cup \{\phi_l\}$ **end if**

Go to **Step 3**

Step 6: if $\mathcal{W} \setminus \cup_{\psi_j \in \mathcal{A}_0} ii(\psi_j) \neq \emptyset$ **then** $ii(true) \leftarrow \mathcal{W} \setminus \cup_{\psi_j \in \mathcal{A}_0} ii(\psi_j)$ **and** $\mathcal{A}_0 \leftarrow \mathcal{A}_0 \cup \{true\}$ **end if**

$ii(false) \leftarrow \emptyset$

$\mathcal{A}_0 \leftarrow \mathcal{A}_0 \cup \{false\}$

end of *algorithm*

[1] Formula ϕ is called a smallest formula in $\mathcal{A}$ if $\forall \psi \in \mathcal{A}, \psi \not\models \phi$.

[2] We assume that $\mathcal{A}$ contains at least one axiom in addition to the tautologies and *false*. Otherwise, this GICT tells nothing apart from that *true* is supported by $\mathcal{W}$ and *false* by empty.

[3] This procedure will terminate because P is finite, so are $\mathcal{L}(P)$ and the set of axioms, $\mathcal{A}$.

In this algorithm and other algorithms, notation $A \leftarrow B$ means that A is assigned with the result of evaluating expression B.

Therefore, function $ii : \mathcal{A}_0 \rightarrow 2^{\mathcal{W}}$ is constructed. $ii(true)$ represents those possible worlds which only support formula $true$.

In Step 4, there may be more than one axiom satisfying the condition. Proposition 3.5 presented below proves the indifference when this happens.

Proposition 3.5:

Applying Algorithm A to a GICT produces the same result regardless of the order of selecting formulae in Step 4.

Proof.

Let $\mathcal{A}_0$ be the set containing axioms with $ii(\psi_j) \neq \emptyset$ and $\mathcal{A}'$ be the set with remaining axioms after Step 3. Assume that there are two formulae ϕ_1, ϕ_2 satisfying the condition specified in Step 4.

When ϕ_1 is chosen first in Step 4, then

$$ii(\phi_1) = \bigcup_{\psi_j \in \mathcal{A}_0, \psi_j \models \phi_1} ii(\psi_j).$$

Then, ϕ_2 is selected with:

$$\begin{aligned} ii(\phi_2) &= i(\phi_2) \setminus \textstyle\bigcup_{\psi_j \in \mathcal{A}_0 \cup \{\phi_1\}, \psi_j \models \phi_2} ii(\psi_j) \\ &= \textstyle\bigcup_{\psi_j \in \mathcal{A}_0, \psi_j \models \phi_2} ii(\psi_j) \text{ (because } \phi_1 \not\models \phi_2), \end{aligned}$$

which indicates that adding ϕ_1 into set $\mathcal{A}_0$ will have no effect on the outcome of $ii(\phi_2)$.

In the same way we can prove that choosing ϕ_2 first and then ϕ_1 gives exactly the same $ii(\phi_1)$ and $ii(\phi_2)$ as above.

Similarly, we can prove the proposition for any set of axioms $\{\phi_1, \ldots, \phi_n\}$ satisfying the condition in Step 4.

QED

Theorem 1 *Given a GICT, $< \mathcal{W}, \mu, P, \mathcal{A}, i >$, function ii constructed from Algorithm A is the unique function satisfying the condition in Definition 3.6.*

Proof.

To make the statement clear, we assume that the function obtained from Algorithm A is ii_0 on set $\mathcal{A}_0$ and the function from Definition 3.6 is ii_1 on set $\mathcal{A}_1$. $\mathcal{A}_0 \subseteq \mathcal{A}$, $\mathcal{A}_1 \subseteq \mathcal{A}$.

Part I: ii_1 is a regulated function on set $\mathcal{A}_0$.

For any $\phi \in \mathcal{A}_0$, we have:

$$\begin{aligned} ii_0(\phi) &= i(\phi) \setminus \bigcup_{\psi_j \in \mathcal{A}_0, \psi_j \models \phi} ii_0(\psi_j) \text{ (equation in Step 4)} \\ &= i(\phi) \setminus \bigcup_{\phi_j \in \mathcal{A}_0, \phi_j \models \phi, \phi_j \neq \phi} (\bigcup_{\psi_{jl} \in \mathcal{A}_0, \psi_{jl} \models \psi_j} ii_0(\psi_{jl}) \cup ii_0(\psi_j)) \\ &\qquad \text{(based on } S \cup S = S\text{, we add some extra } ii_0(\psi_{jl}) \text{ sets)} \\ &= i(\phi) \setminus \bigcup_{\phi_j \in \mathcal{A}_0, \phi_j \models \phi, \phi_j \neq \phi} i(\psi_j) \\ &\qquad \text{(through the reverse of the equation in Step 4)} \\ &= ii_1(\phi) \text{ (from Definition 3.5).} \end{aligned}$$

ii_0 is equivalent to ii_1 on set $\mathcal{A}_0$.

Furthermore, $\forall \phi \in \mathcal{A}_1, ii_1(\phi) \neq \emptyset$ when $\phi \neq false$, ii_0 is a regulated function on $\mathcal{A}_0$.

Part II: Proof that ii_0 is identical to ii_1.

The proof of Part I shows that $\mathcal{A}_0 \subseteq \mathcal{A}_1$. We only need to prove that $\mathcal{A}_1 \subseteq \mathcal{A}_0$.

At the beginning of Algorithm A, let the smallest set of axioms from formula (3.5) be in $\mathcal{A}_0'$. Then, $\mathcal{A}_0' \subseteq \mathcal{A}_1$, and $ii_0(\phi) = ii_1(\phi)$ for every $\phi \in \mathcal{A}_0'$.

We choose an axiom $\psi \in \mathcal{A}_1 \setminus \mathcal{A}_0'$ where for any other $\psi_j \in \mathcal{A}_1 \setminus \mathcal{A}_0'$, if $\psi_j \neq \psi$, then $\psi_j \not\models \psi$. Then, there is a list of $\psi_l \in \mathcal{A}_0$ where $\psi_l \models \psi$; otherwise, ψ in $\mathcal{A}_0'$.

Based on Step 4 in Algorithm A, we have:

$$\begin{aligned} ii_0(\psi) &= i(\psi) \setminus \bigcup_{\psi_l \in \mathcal{A}_0', \psi_l \models \psi} ii_0(\psi_l) \\ &= i(\psi) \setminus \bigcup_{\psi_l \in \mathcal{A}_0', \psi_l \models \psi} ii_1(\psi_l) \text{ (as } ii_0(\psi_l) = ii_1(\psi_l)) \\ &\supseteq i(\psi) \setminus \bigcup_{\psi_l \in \mathcal{A}_0', \psi_l \models \psi} i(\psi_l) \text{ (using } i(\psi) \text{ to replace } ii(\psi)) \end{aligned}$$

As $ii_1(\psi) = i(\psi) \setminus \bigcup_{\psi_l \in \mathcal{A}_0', \psi_l \models \psi} i(\psi_l)$ is not empty ($\psi \in \mathcal{A}_1$), so is $ii_0(\psi)$. Therefore, ψ is in $\mathcal{A}_0$.

Now, let us redefine $\mathcal{A}_0' = \mathcal{A}_0' \cup \{\psi\}$, and repeat the above procedure for another axiom ψ_l in $\mathcal{A}_1 \setminus \mathcal{A}_0'$. We will prove again that ψ_l is in $\mathcal{A}_0$. Eventually, every formula which is in $\mathcal{A}_1$ is proved to be in $\mathcal{A}_0$. That is, $\mathcal{A}_1 \subseteq \mathcal{A}_0$. Therefore, $\mathcal{A}_0$ and $\mathcal{A}_1$ are the same set and ii_0 and ii_1 are identical on this set.

Part III: Proof that ii_1 is the unique function of this kind.

This is a natural consequence of Proposition 3.5.

QED

Proposition 3.6:

Let ϕ_1 and ϕ_2 be two axioms in $\mathcal{A}_0$ at any time in Algorithm A, then

$$ii(\phi_1) \cap ii(\phi_2) = \emptyset.$$

Proof.

We choose two axioms ϕ_1 and ϕ_2 randomly from $\mathcal{A}_0$:

1. When $\phi_1 \models \phi_2$, then $ii(\phi_2) = i(\phi_2) \setminus (ii(\phi_1) \bigcup_{\psi_l \in \mathcal{A}_0, \psi \neq \phi_1 \psi_l \models \phi_2} ii_1(\psi_l))$. Therefore, $ii(\phi_1) \cap ii(\phi_2) = \emptyset$. As ϕ_1 and ϕ_2 are selected arbitrarily, this is true when $\phi_2 \models \phi_1$.

2. When $\phi_1 \not\models \phi_2$ and $\phi_2 \not\models \phi_1$, assume $ii(\phi_1) \cap ii(\phi_2) \neq \emptyset$, there is at least one possible world w such that $w \in ii(\phi_1) \cap ii(\phi_2)$. We have:

$$\begin{aligned}
& w \in ii(\phi_1) \cap ii(\phi_2) \\
& \Rightarrow w \in i(\phi_1) \cap i(\phi_2) = i(\phi_1 \wedge \phi_2) \\
& \Rightarrow \exists \psi, w \in i(\psi), \psi = \phi_1 \wedge \phi_2, \psi \neq false \\
& \Rightarrow w \in ii(\psi) \text{ or } w \in ii(\psi_l) \text{ such that } \psi_l \models \psi \\
& \Rightarrow w \notin ii(\phi_1), w \notin ii(\phi_2) \text{ based on Step 4} \\
& \quad\ (\text{remember } \psi \models \phi_1, \psi_l \models \phi_2, \psi_l \models \phi_1, \psi \models \phi_2) \\
& \Rightarrow w \notin ii(\phi_1) \cap ii(\phi_2).
\end{aligned}$$

   Tis is contradictory. Therefore, $ii(\phi_1) \cap ii(\phi_2) = \emptyset$.

QED

Here, $\psi \neq false$ means that ψ is not invalid. This special property of function ii promoted us with the following definition.

Definition 3.7: *Basic Incidence Assignment*

Given a set of axioms $\mathcal{A}_0$ and a set of possible worlds $\mathcal{W}$, function $ii : \mathcal{A}_0 \to 2^{\mathcal{W}}$ is called a basic incidence assignment if ii satisfies the following conditions:

$$\begin{array}{ll}
ii(false) = \emptyset, & false \in \mathcal{A}_0, \\
ii(\phi) \neq \emptyset, & \forall \phi \in \mathcal{A}_0, \phi \neq false, \\
ii(\phi) \cap ii(\psi) = \emptyset, & when\ \phi, \psi \in \mathcal{A}_0, \phi \neq \psi\ (\phi \not\models \psi\ or\ \psi \not\models \phi), \\
\bigcup_{\phi_j} ii(\phi_j) = \mathcal{W}, & \phi_j \in \mathcal{A}_0.
\end{array}$$

In a basic incidence assignment, each possible world w is assigned to one and only one specific formula. When $w \in ii(\phi_l)$, then ϕ_l is the smallest formula that w supports (by smallest, we mean that if w also supports ϕ_j, then $\phi_l \models \phi_j$ $(\phi_j \neq \phi_l)$).

Therefore, a basic incidence assignment gives a unique mapping relation between each possible world and the smallest formula of all the formulae it supports. However, an incidence function i maps a possible world w to every formula it supports. For instance, if formula ϕ is true under w, then formulae such as $\phi \vee \psi_1$, $\phi \vee \psi_2$, ..., etc. are all true under w.

In a basic incidence assignment, w only belongs to $ii(\phi)$, and $w \notin ii(\phi \vee \psi_l)$, however, for an incidence function i, $w \in i(\phi)$ and $w \in i(\phi \vee \psi_l)$. Therefore, a basic incidence assignment is more fundamental than an incidence function.

Proposition 3.7:

Any regulated function ii from an incidence function i in a given GICT is a basic incidence assignment.

Proof.

If ii has been regulated on domain $\mathcal{A}_0$, then $false \in \mathcal{A}_0$ and $ii(false) = \emptyset$.

Based on Proposition 3.6, $ii(\phi_i) \cap ii(\phi_j) = \emptyset$ holds for any two distinct elements ϕ_i and ϕ_j in $\mathcal{A}_0$.

According to Algorithm A, $\forall \phi \in \mathcal{A}_0, \phi \neq false$, $ii(\phi) \neq \emptyset$. $ii(true) = \mathcal{W} \setminus \cup_{\phi_l \in \mathcal{A}_0} ii(\phi_l)$ when $ii(true) \neq \emptyset$, so that $\cup_{\phi_l \in \mathcal{A}_0} ii(\phi_l) = \mathcal{W}$.

ii is a basic incidence assignment.

QED

Example 3.4

Following the incidence assignment in Example 2.1, the GICT is

$$< \mathcal{W}, \mu, P, \mathcal{A}, i >$$

where $\mathcal{W}$ contains seven possible worlds, and $\mu(w) = 1/7$, $P = \{sunny$, $windy$, $rainy\}$ and $\mathcal{A} = \{rainy, windy, rainy \wedge windy, true, false\}$[4]. The incidence function assigns the following incidence sets to the axioms in $\mathcal{A}$:

$$\begin{aligned} &i(rainy) = \{fri, sat, sun, mon\}, \\ &i(windy) = \{mon, wed, fri\}, \\ &i(rainy \wedge windy) = \{mon, fri\}, \\ &i(true) = \mathcal{W} \text{ and } i(false) = \emptyset. \end{aligned}$$

[4]It is worth noting that the set of axioms $\mathcal{A}$ in this GICT is closed under $\wedge$ and is different from that in the original incidence calculus theory in Example 2.1 (the original incidence calculus does not require this condition).

By Definition 3.5, function ii on set $\mathcal{A}$ can then be defined as:

$$ii(rainy \wedge windy) = \{fri, mon\},$$
$$ii(rainy) = \{sat, sun\},\ ii(windy) = \{wed\},$$
$$ii(true) = \{tues, thur\}, ii(false) = \emptyset.$$

This ii is a regulated one and is a basic incidence assignment on set $\mathcal{A}$, according to Definitions 3.6 and 3.7.

The incidence function can be recovered from ii as:

$$i(rainy \wedge windy) = ii(rainy \wedge windy) = \{mon, fri\},$$
$$i(rainy) = ii(rainy) \cup ii(rainy \wedge windy) = \{fri, sat, sun, mon\},$$
$$i(windy) = ii(windy) \cup ii(rainy \wedge windy) = \{mon, wed, fri\},$$
$$i(true) = ii(true) \cup ii(rainy) \cup ii(windy) \cup ii(rainy \wedge windy) = \mathcal{W}.$$

sat in $ii(rainy)$ means that the smallest axiom that this possible world supports is $rainy$. While the smallest axiom that fri supports is $rainy \wedge windy$ although $fri \in i(rainy)$ as well.

This example demonstrates the relationship between an incidence function and a basic incidence assignment. That is, one function can be derived from another. In the next section, we study this phenomenon in depth to see their formal relationship.

3.3 An Incidence Function Has a Unique Basic Incidence Assignment

Theorem 2 given below proves that function ii obtained from Algorithm A is a basic incidence assignment and the corresponding incidence function can be recovered from it.

Therefore, given an incidence function i, there is always a unique basic incidence assignment from which it can be recovered.

Now, we have the following theorem.

Theorem 2 *Given a GICT, $< \mathcal{W}, \mu, P, \mathcal{A}, i >$, function $ii : \mathcal{A}_0 \to 2^{\mathcal{W}}$ obtained from Algorithm A is the unique basic incidence assignment derivable from function i in this GICT and i can be recovered from it using equation (3.6) as*

$$i(\phi) = \bigcup_{\phi_j \in \mathcal{A}_0, \phi_j \models \phi} ii(\phi_j). \tag{3.6}$$

Proof.

Part I: ii is a basic incidence assignment.

This is a natural consequence of Proposition 3.7.

Part II: i can be recovered from ii using equation (3.6).

For each formula $\phi \in \mathcal{A}$, if ϕ is also in $\mathcal{A}_0$, then there are two situations when ϕ is added into $\mathcal{A}_0$ in Algorithm A:

1. ϕ is added into $\mathcal{A}_0$ at Step 1, where

$$i(\phi) = ii(\phi) = \bigcup_{\phi_j \in \mathcal{A}_0, \phi_j \models \phi} ii(\phi_j).$$

2. ϕ is added into $\mathcal{A}_0$ in Steps 4 and 5, where

$$ii(\phi) = i(\phi) \setminus (\bigcup_{\phi_j \in \mathcal{A}_0, \phi_j \models \phi} ii(\phi_j))$$

   and

$$ii(\phi) \neq \emptyset,$$

   so that

$$i(\phi) = ii(\phi) \cup (\bigcup_{\phi_j \in \mathcal{A}_0, \phi_j \models \phi} ii(\phi_j)) = \bigcup_{\phi_j \in \mathcal{A}_0, \phi_j \models \phi} ii(\phi_j).$$

For each formula $\phi \in \mathcal{A} \setminus \mathcal{A}_0$, in Step 4

$$ii(\phi) = i(\phi) \setminus (\bigcup_{\phi_j \in \mathcal{A}_0, \phi_j \models \phi} ii(\phi_j)),$$

and in Step 5 $ii(\phi) = \emptyset$, so that the equation

$$i(\phi) = \bigcup_{\phi_j \in \mathcal{A}_0, \phi_j \models \phi} ii(\phi_j)$$

holds.

Therefore, no matter in which case, the incidence set $i(\phi)$ of ϕ can always be recovered from ii using equation (3.6).

Part III: ii is unique.

This can easily be proved from Theorem 1.

QED

Example 3.5 (a simplified form of an example from [Chateauneuf, 1994])

Suppose there are four urns of balls. The balls in the first urn are all *Blue*, the balls in the second urn are either *Blue* or *Green*, the balls in the third urn are all *Red* and the balls in the fourth urn are either *Green* or *Red*.

Suppose one ball is drawn from one of the four urns randomly (there is no information on which urn is preferable) and we are interested in the colour of the drawing ball.

Let P be a set of atomic propositions consisting of the following three propositions:

- q_1: the ball is *Blue*,
- q_2: the ball is *Green*,
- q_3: the ball is *Red*.

It is possible to establish the supporting relations between the event "drawing a ball from an urn" and the propositions in P using the incidence function i:

$$\begin{array}{ll} i(q_1) = \{1\}, & i(q_1 \vee q_2) = \{1, 2\}, \\ i(q_3) = \{3\}, & i(q_2 \vee q_3) = \{3, 4\}, \\ i(true) = \{1, 2, 3, 4\}, & i(false) = \emptyset, \end{array}$$

where $1, 2, 3, 4$ stand for four possible worlds in which a ball is drawn from urns 1, 2, 3 and 4, respectively.

As $i((q_1 \vee q_2) \wedge (q_2 \vee q_3)) = i(q_1 \vee q_2) \cap i(q_2 \vee q_3) = \emptyset$, the set of axioms $\mathcal{A}$ actually is $\mathcal{A} = \{q_1, q_2, q_3, q_4, true, false\}$ which is closed under $\wedge$ and the corresponding GICT is:

$$< \mathcal{W}, \mu, P, \mathcal{A}, i >,$$

where $\mathcal{W} = \{1, 2, 3, 4\}$ and $\mu(w) = 1/4$.

A regulated function ii on set $\mathcal{A} \backslash \{true\}$ can be defined through Definitions 3.5 and 3.6 and it is the basic incidence assignment:

$$\begin{array}{ll} ii(q_1) = \{1\}, & ii(q_1 \vee q_2) = \{2\}, \\ ii(q_3) = \{3\}, & ii(q_2 \vee q_3) = \{4\}, \\ ii(false) = \emptyset. & \end{array}$$

The incidence function i on $\mathcal{A}$ can be recovered from ii as:

$$i(q_1) = ii(q_1), \quad i(q_1 \vee q_2) = ii(q_1) \cup ii(q_1 \vee q_2),$$
$$i(q_3) = ii(q_3), \quad i(q_2 \vee q_3) = 278ii(q_3) \cup ii(q_2 \vee q_3),$$
$$i(true) = \bigcup_j ii(q_j) = \mathcal{W} \quad i(false) = \emptyset.$$

Therefore, the verdict about the colour of the drawing ball is: its colour is blue with probability 0.25, is blue or red with 0.5, is red with 0.25 and is green or red with 0.5.

$\diamondsuit$

Recovering an incidence function i from ii is not difficult, if incidence function i is given initially and ii is derived from it. However, if a basic incidence assignment ii on domain $\mathcal{A}_0$ is given first, what is the prospect of defining an incidence function i?

So, is there any method of finding a domain $\mathcal{A}$, which should be closed under $\wedge$, on which i is defined?

3.4 A Basic Incidence Assignment Maps to a Family of Incidence Assignments

In Section 3.3, we focused on how to obtain a basic incidence assignment, ii, from an incidence function.

In this section, we look at this problem the other way round, i.e., given a basic incidence assignment ii on $\mathcal{A}_0$, what is the approach to finding a suitable set $\mathcal{A}$ on which i can be defined from ii? Also, is this set $\mathcal{A}$ unique? If not, how can we enumerate them?

In this section, we will examine these problems in detail.

Example 3.6

Given a GICT, let its incidence function assign incidences on axioms in $\mathcal{A}$ as follows:

$$i(rainy) = \{fri, sat, sun, mon\},$$
$$i(windy) = \{mon, wed, fri\},$$
$$i(rainy \wedge windy) = \{mon, fri\},$$
$$i(windy \vee sunny) = \{mon, wed, fri\},$$
$$i(rainy \wedge (windy \vee sunny)) = \{mon, fri\},$$
$$i(true) = \mathcal{W} \text{ and } i(false) = \emptyset,$$

where $\mathcal{A} = \{windy, rainy, rainy \wedge windy, windy \vee sunny, rainy \wedge (windy \vee sunny), true, false\}$.

A basic incidence assignment, ii, can be defined on domain $\mathcal{A}_0$=$\{windy \wedge rainy, windy, rainy, true, false\}$ as:

$$\begin{aligned}
&ii(rainy \wedge windy) = \{fri, mon\},\\
&ii(rainy) = \{sat, sun\},\\
&ii(windy) = \{wed\},\\
&ii(true) = \{tues, thur\},\\
&ii(false) = \emptyset.
\end{aligned}$$

This GICT shares the same basic incidence assignment with the GICT in Example 3.4, but contains more axioms than that in Example 3.4. The set of axioms in Example 3.4 is a proper subset of the set of axioms in Example 3.6. This suggests that a family of incidence functions, defined on different sets of axioms, can be mapped to the same basic incidence assignment.

Therefore, in order to derive incidence functions from basic incidence assignments, we will need to answer at least the following two questions:

1. Can we find that incidence function, which is defined on the smallest set of axioms in a family of sets of axioms, given an ii on $\mathcal{A}_0$ initially?

2. Can we generate other incidence functions from this special one on other sets of axioms in this family, and guarantee that all these incidence functions share the same ii that is given in the first place?

The two theorems given below provide solutions for these two questions, respectively.

Theorem 3 *Given a set of axioms $\mathcal{A}_0$ with a basic incidence assignment $ii : \mathcal{A}_0 \to 2^{\mathcal{W}}$, let*

$$\mathcal{A} = \mathcal{A}_0 \cup \{true, false\} \cup \{\phi \mid \phi = \psi_1 \wedge \ldots \wedge \psi_n, \psi_1, \ldots, \psi_n \in \mathcal{A}_0, n > 1\}.$$

Then, $\mathcal{A}$ is the smallest set of axioms containing $\mathcal{A}_0$ and is closed under $\wedge$.

Function i defined on $\mathcal{A}$ by equation (3.6) is an incidence function.

Proof.

Part I: $\mathcal{A}$ is the smallest set among all possible sets of axioms containing $\mathcal{A}_0$ and closed under $\wedge$.

It is easy to see from the way that $\mathcal{A}$ is constructed that $\mathcal{A}$ is the smallest set which contains $\mathcal{A}_0$ and closed under $\wedge$.

Part II: i is an incidence assignment on $\mathcal{A}$.

First of all, $false \in \mathcal{A}_0$, so $false \in \mathcal{A}$. Define $i(false) = ii(false) = \emptyset$.
Secondly, when $true \in \mathcal{A}_0$, $ii(true) \neq \emptyset$. $ii(true) = \mathcal{W} \setminus \cup_j ii(\phi_j)$, so that $i(true) = ii(true) \cup (\cup_j ii(\phi_j)) = \mathcal{W}$. Otherwise, define $i(true) = \cup_j ii(\phi_j) = \mathcal{W}$.
Thirdly, as $\mathcal{A}$ is proved closed under $\wedge$, we only need to prove that $i(\phi \wedge \psi) = i(\phi) \cap i(\psi)$ when ϕ, ψ and $\phi \wedge \psi$ are all in $\mathcal{A}$.
Suppose that $i(\phi) \cap i(\psi) = \mathcal{W}' \neq \emptyset$, and:

$$\begin{array}{ll} \forall w \in \mathcal{W}', & w \in i(\phi) \cap i(\psi) \\ \Leftrightarrow \exists \phi_0, & w \in ii(\phi_0) \quad (\phi_0 \models \phi,\ \phi_0 \models \psi) \\ \Leftrightarrow \exists \phi_0, & w \in ii(\phi_0), \quad \phi_0 \models \phi \wedge \psi \\ \Leftrightarrow \exists \phi_0, & w \in ii(\phi_0) \wedge ii(\phi_0) \subseteq i(\phi \wedge \psi) \text{ (from equation (3.6))} \\ \Leftrightarrow & w \in i(\phi \wedge \psi) \end{array}$$

So, $i(\phi) \wedge i(\psi) = i(\phi \wedge \psi)$ holds.
When $i(\phi) \wedge i(\psi) = \emptyset$, then it is also easy to prove that $i(\phi) \wedge i(\psi) = i(\phi \wedge \psi)$.
Therefore, function i defined by (3.6) is an incidence function.

QED

Theorem 4 *Let ii be a basic incidence assignment $ii : \mathcal{A}_0 \rightarrow 2^{\mathcal{W}}$, and $\mathcal{A}$ be the smallest set of axioms containing $\mathcal{A}_0$ defined in Theorem 3.*
Then, set $\mathcal{B}$ defined by equation (3.7) shown below:

$$\mathcal{B} = \mathcal{A} \cup \{\phi \mid \phi = \psi \vee \phi', \phi' \in \mathcal{A}\}, \tag{3.7}$$

where $\psi \in \mathcal{L}(P)$ but $\psi \notin \mathcal{A}$, is a member of a family of sets of axioms, on which there is an incidence function, $i_{\mathcal{B}}$, that has ii as its basic incidence assignment.

Proof.

We need to prove three separate results for this theorem. First of all, $\mathcal{B}$ is closed under $\wedge$. Secondly, there is an incidence function $i_{\mathcal{B}}$ on domain $\mathcal{B}$. Thirdly, the basic incidence assignment of $i_{\mathcal{B}}$ is exactly the same as ii.

Part I: $\mathcal{B}$ is closed under $\wedge$.

Assume that ϕ_1 and ϕ_2 are two distinct formulae in $\mathcal{B}$.
Then:

(i) If both of these formulae are in $\mathcal{A}$, then $\phi_1 \wedge \phi_2 \in \mathcal{A} \subseteq \mathcal{B}$. $\mathcal{B}$ is closed.

(ii) If both of these formulae are in $\mathcal{B} \setminus \mathcal{A}$, then:

$$\begin{aligned}
&\phi_1, \phi_2 \in \mathcal{B} \\
&\Leftrightarrow \exists \phi_1' \in \mathcal{A}, \exists \phi_2' \in \mathcal{A}, \phi_1 = \phi_1' \vee \psi, \phi_2 = \phi_2' \vee \psi \\
&\Leftrightarrow \exists \phi_1' \in \mathcal{A}, \exists \phi_2' \in \mathcal{A}, \phi_1 \wedge \phi_2 = (\phi_1' \vee \psi) \wedge (\phi_2' \vee \psi) \\
&\Leftrightarrow \exists \phi_1' \in \mathcal{A}, \exists \phi_2' \in \mathcal{A}, \phi_1 \wedge \phi_2 = (\phi_1' \wedge \phi_2') \vee \psi \\
&\Leftrightarrow \exists \phi'' \in \mathcal{A}, \phi'' = \phi_1' \wedge \phi_2', \phi_1 \wedge \phi_2 = \phi'' \vee \psi \\
&\Leftrightarrow \phi_1 \wedge \phi_2 \in \mathcal{B} \text{ (because } \phi'' \vee \psi \in \mathcal{B} \text{ according to equation(3.7))}
\end{aligned}$$

So, $\mathcal{B}$ is closed under $\wedge$.

(iii) If one of these formulae is in $\mathcal{B} \setminus \mathcal{A}$, then we can use the similar approach as in (ii) to prove that $\mathcal{B}$ is closed under $\wedge$.

Therefore, the set constructed using (3.7) is closed under $\wedge$.

Part II: There exists an incidence function $i_{\mathcal{B}}$ on domain $\mathcal{B}$.

Function $i_{\mathcal{B}}$ is defined on $\mathcal{B}$ as:

$$i_{\mathcal{B}}(\phi) = \begin{cases} i_{\mathcal{A}}(\phi) & \text{if } \phi \in \mathcal{A} \\ \bigcup_{\phi_j \in \mathcal{A}, \phi_j \models \phi} i_{\mathcal{A}}(\phi_j) & \text{otherwise,} \end{cases} \tag{3.8}$$

where $i_{\mathcal{A}}$ is the incidence function on $\mathcal{A}$ obtained from Theorem 3.

Assume that ϕ_1 and ϕ_2 are two distinct formulae in $\mathcal{B}$, we need to prove that $i_{\mathcal{B}}(\phi_1) \wedge i_{\mathcal{B}}(\phi_2) = i_{\mathcal{B}}(\phi_1 \wedge \phi_2)$.

We have:

(i) When both of these formulae are in $\mathcal{A}$, $\phi_1 \wedge \phi_2$ is also in $\mathcal{A}$. We have $i_{\mathcal{B}}(\phi_1 \wedge \phi_2) = i_{\mathcal{A}}(\phi_1 \wedge \phi_2) = i_{\mathcal{A}}(\phi_1) \cap i_{\mathcal{A}}(\phi_2) = i_{\mathcal{B}}(\phi_1) \cap i_{\mathcal{B}}(\phi_2)$.

(ii) When both of these formulae, ϕ_1, ϕ_2, are in $\mathcal{B} \setminus \mathcal{A}$, assume that $i_{\mathcal{B}}(\phi_1) \cap i_{\mathcal{B}}(\phi_2) = \mathcal{W}' \neq \emptyset$.

Then:

$$\begin{aligned}
&\forall w \in \mathcal{W}' \\
&\Leftrightarrow w \in i_{\mathcal{B}}(\phi_1) \cap i_{\mathcal{B}}(\phi_2) \\
&\Leftrightarrow \exists \phi_1', \phi_2' \in \mathcal{A}, w \in i_{\mathcal{B}}(\phi_1) \cap i_{\mathcal{B}}(\phi_2), \phi_1 = \phi_1' \vee \psi, \phi_2 = \phi_2' \vee \psi \\
&\Leftrightarrow \exists \phi_0,\ w \in ii(\phi_0) \quad \phi_0 \models \phi_1',\ \phi_0 \models \phi_2', \phi_0 \in \mathcal{A}_0, \phi_1 = \phi_1' \vee \psi, \phi_2 = \phi_2' \vee \psi \\
&\Leftrightarrow \exists \phi_0,\ w \in ii(\phi_0), \quad \phi_0 \models \phi_1' \wedge \phi_2', \phi_1 = \phi_1' \vee \psi, \phi_2 = \phi_2' \vee \psi \\
&\Leftrightarrow \exists \phi_0,\ w \in i_{\mathcal{A}}(\phi_1' \wedge \phi_2'), \phi_1 \wedge \phi_2 = (\phi_1' \wedge \phi_2') \vee \psi \\
&\Leftrightarrow w \in i_{\mathcal{B}}(\phi_1 \wedge \phi_2), \text{ as } i_{\mathcal{A}}(\phi_1' \wedge \phi_2') \subseteq i_{\mathcal{B}}((\phi_1' \wedge \phi_2') \vee \psi) = i_{\mathcal{B}}(\phi_1 \wedge \phi_2).
\end{aligned}$$

$i_{\mathcal{B}}(\phi_1) \wedge i_{\mathcal{B}}(\phi_2) = i_{\mathcal{B}}(\phi_1 \wedge \phi_2)$.

(iii) Similarly, when only one of these formulae is in $\mathcal{B}$, we can still prove that $i_{\mathcal{B}}(\phi_1) \wedge i_{\mathcal{B}}(\phi_2) = i_{\mathcal{B}}(\phi_1 \wedge \phi_2)$.

Because *true* and *false* are axioms in $\mathcal{A}$, and $i_{\mathcal{B}}(true) = i_{\mathcal{A}}(true) = \mathcal{W}$ and $i_{\mathcal{B}}(false) = i_{\mathcal{A}}(false) = \emptyset$, therefore, $i_{\mathcal{B}}$ is an incidence function on $\mathcal{B}$.

Part III: The basic incidence assignment of $i_{\mathcal{B}}$ is ii on domain $\mathcal{A}_0$.

To prove that ii is a basic incidence assignment of $i_{\mathcal{B}}$, we need to show that $ii_{\mathcal{B}}(\phi) = \emptyset$ when $\phi \notin \mathcal{A}$ and $ii_{\mathcal{B}}(\phi) = ii(\phi)$ when $\phi \in \mathcal{A}$.

For any $\phi \in \mathcal{B} \setminus \mathcal{A}$, we have:

$$\begin{aligned}
ii_{\mathcal{B}}(\phi) &= i_{\mathcal{B}}(\phi) \setminus \textstyle\bigcup_{\phi_j \in \mathcal{B}, \phi_j \models \phi} i_{\mathcal{B}}(\phi_j) \text{ (from Definition 3.5)} \\
&= \textstyle\bigcup_{\phi_t \in \mathcal{A}, \phi_t \models \phi} i_{\mathcal{A}}(\phi_t) \setminus \bigcup_{\phi_j \in \mathcal{B}, \phi_j \models \phi} i_{\mathcal{B}}(\phi_j) \\
&\qquad \text{(replace } i_{\mathcal{B}}(\phi) \text{ using equation (3.8))} \\
&= \textstyle\bigcup_{\phi_t \in \mathcal{A}, \phi_t \models \phi} i_{\mathcal{A}}(\phi_t) \setminus \bigcup_{\phi_j \in \mathcal{B}, \phi_j \models \phi} (\bigcup_{\phi_j \in \mathcal{A}, \phi_l \models \phi_j} i_{\mathcal{A}}(\phi_l)) \\
&\qquad \text{(replace each } i_{\mathcal{B}}(\phi_j) \text{ using equation (3.8))} \\
&= \textstyle\bigcup_{\phi_t \in \mathcal{A}, \phi_t \models \phi} i_{\mathcal{A}}(\phi_t) \setminus \bigcup_{\phi_l \in \mathcal{A}, \phi_l \models \phi} i_{\mathcal{A}}(\phi_l) \\
&\qquad \text{(as } \phi_j \models \phi, \phi_l \models \phi_j, \text{ so } \phi_l \models \phi) \\
&= \emptyset \text{ (as } \{\phi_t \mid \phi_t \in \mathcal{A}, \phi_t \models \phi\} = \{\phi_l \mid \phi_l \in \mathcal{A}, \phi_l \models \phi\}).
\end{aligned}$$

For $\phi \in \mathcal{A}$, we have:

$$\begin{aligned}
ii_{\mathcal{B}}(\phi) &= i_{\mathcal{B}}(\phi) \setminus \textstyle\bigcup_{\phi_j \in \mathcal{B}, \phi_j \models \phi} i_{\mathcal{B}}(\phi_j) \\
&= i_{\mathcal{A}}(\phi) \setminus \textstyle\bigcup_{\phi_j \in \mathcal{A}, \phi_j \models \phi} i_{\mathcal{A}}(\phi_j) \\
&= ii(\phi)
\end{aligned}$$

Therefore, the basic incidence assignment derived from $i_{\mathcal{B}}$ is identical to ii.

QED

The conclusion we draw from the above three theorems is that a single incidence assignment is mapped to a unique basic incidence assignment. However, a single basic incidence assignment can be mapped to a family of incidence assignments.

3.5 Summary

In this chapter, we generalized the original incidence calculus to make incidence calculus capable of representing a wider range of information.

The incidence function i is recognised as playing a dual role: on the one hand, it is the bridge between symbolic and numerical reasoning mechanisms which makes incidence calculus different; on the other hand, its truth functionality reduces its expressive power in reality.

As a result, the underlying propositional logic that supports the definition of incidence functions is modified from two-valued to three-valued to represent incomplete information (ignorance).

Generalized incidence calculus is thus ready to represent information beyond the scope of the original incidence calculus.

An important function, basic incidence assignment, emerged in this chapter, to demonstrate the nature of mapping relationships between possible worlds and formulae.

In a basic incidence assignment, each possible world is attached to only one formula, but in an incidence function, the same possible world is attached to all the formulae which are true in it.

Therefore, basic incidence assignments are more fundamental than incidence functions.

In summary, each incidence function has only one basic incidence assignment and one basic incidence assignment is consistent with a family of incidence functions. In the context of numerical theories, such as DS theory, the basic incidence assignment is equivalent to the concept of mass function, and the incidence function is equivalent to the belief function.

It is worth noting that when a set of atomic propositions involved in a GICT is very large, it is very inefficient to infer lower and upper bounds of incidence of formulae, given an incidence function. In [Liu 1996], the benefit of splitting a large set of propositions into a number of smaller ones is discussed. It was shown that using many different but inter-related smaller sets of propositions is much more efficient in inference than using just a larger one.

Chapter 4

From Numerical to Symbolic Assignments

Given a generalized incidence calculus theory (GICT), the lower bounds of probabilities on formulae can be inferred from the lower bounds of incidence sets using equation (3.2), However sometimes numerical assignments are given on some formulae directly without defining any incidence functions. When this happens, incidence calculus cannot be applied at all. This is its second limitation.

In Chapter 2, two methods were given for converting a numerical assignment into a consistent incidence assignment based on a tree structure. These methods can also be applied in generalized incidence calculus. However, the limitations of these methods are heavy rocks on the way of applying them, in particular, related to an enormous time wasted on searching *all* consistent assignments.

In this chapter, we aim at designing a faster algorithm to obtain consistent incidence functions. The ideal solution is that a limited number of consistent assignments would be enough to subsume all possible ones. The best situation is that only one assignment is essential.

As there is only one basic incidence assignment for an incidence function, and this basic incidence assignment can be mapped to a family of such incidence functions, the best way to recover a symbolic assignment is to search for a basic incidence assignment first, and then to generate all the necessary consistent incidence functions. So discovering a basic incidence assignment from a numerical assignment is the crucial step in the whole recovery procedure.

4.1 An Algorithm for Assigning Incidences

In this section, we discuss how to extend Algorithm A to find a basic incidence assignment from a numerical assignment. We first abstract out a function, *ii*, then prove that *ii* is a basic incidence assignment.

Algorithm B: From Numerical Assignments to Incidence Functions:

Input: *a set* $\mathcal{A}$ *(closed under* $\wedge$*) and* $Prob_*$ *on* $\mathcal{A}$.
Output: *a set* $\mathcal{W}$ *with probability distribution* μ*, a function* $ii : \mathcal{A}_0 \rightarrow 2^{\mathcal{W}}$ *(*$\mathcal{A}_0 \subseteq \mathcal{A}$*), and a function* $i :: \mathcal{A} \rightarrow 2^{\mathcal{W}}$.

Step 1: choose *a subset* $\mathcal{A}_0 = \{\psi_1, \ldots, \psi_n\}$ *of* $\mathcal{A} \setminus \{false\}$*, where* $\mathcal{A}_0$ *satisfies the condition* $\forall \psi_i \in \mathcal{A}_0$, $\forall \phi \in \mathcal{A}$, $\phi \not\models \psi_i$ *if* $\phi \neq \psi_i$

(see **Step 0** *in Algorithm A for details)*

$\mathcal{W} \leftarrow \emptyset$

for each $\phi_j \in \mathcal{A}_0$ **do**

if $Prob_*(\phi_j) > 0$ **then**

$$
\begin{aligned}
&\mathcal{W} \leftarrow \mathcal{W} \cup \{w_j\} \\
&Prob'_*(w_j) = Prob_*(\phi_j), \\
&\mu(w_j) = Prob'_*(w_j), \\
&ii(\phi_j) = \{w_j\}, \\
&\mathcal{A}' = \mathcal{A} \setminus (\mathcal{A}_0 \cup \{false\})
\end{aligned}
$$

end if

$l \leftarrow \mid \mathcal{A}_0 \mid$

Step 2: if $\mathcal{A}'$ *is left with only tautologies* **then** *go to* **Step 4**

Step 3: choose *a formula* ϕ_l *in* $\mathcal{A}'$ **such that**

for each $\phi_j \in \mathcal{A}'$ **if** $\phi_j \neq \phi_l$, **then** $\phi_j \not\models \phi_l$

$Prob'_*(\psi) \leftarrow Prob_*(\psi) - \Sigma_{\phi_j \in \mathcal{A}_0, \phi_j \models \psi} Prob'_*(\phi_j)$.

if $Prob'_*(\psi) > 0$ **then**

$$
\begin{aligned}
&\mathcal{W} \leftarrow \mathcal{W} \cup \{w_{l+1}\}, \\
&\mu(w_{l+1}) \leftarrow Prob'_*(\psi), \\
&ii(\psi) \leftarrow \{w_{l+1}\}, \\
&\mathcal{A}_0 \leftarrow \mathcal{A}_0 \cup \{\psi\}, \\
&\mathcal{A}' \leftarrow \mathcal{A}' \setminus \{\psi\}, \\
&l \leftarrow l + 1
\end{aligned}
$$

else if $Prob'_*(\psi) = 0$ **then** $\mathcal{A}' \leftarrow \mathcal{A}' \setminus \{\psi\}$

else if $Prob'_*(\psi) < 0$ **then** *this assignment is not consistent, stop the procedure*

end if

go to **Step 2**

Step 4: if $\Sigma_j(Prob'_*(\phi_j)) < 1$ **then**

$$\begin{aligned}&\mathcal{W} \leftarrow \mathcal{W} \cup \{w_{l+1}\}\\ &\mu(w_{l+1}) \leftarrow 1 - \Sigma_j Prob'_*(\phi_j),\\ &\mathcal{A}_0 \leftarrow \mathcal{A}_0 \cup \{true\},\\ &ii(true) \leftarrow \{w_{l+1}\}\end{aligned}$$

end if

Step 5: $\mathcal{A}_0 \leftarrow \mathcal{A}_0 \cup \{false\}$
$ii(false) \leftarrow \emptyset$
for each $\phi \in \mathcal{A}$ **do** $i(\phi) \leftarrow \bigcup_{\phi_j \in \mathcal{A}_0, \phi_j \models \phi} ii(\phi_j)$
$i(false) \leftarrow \emptyset$

end of *algorithm*

If there are n elements in $\mathcal{A}$ then there are at most $n+1$ elements in $\mathcal{W}$.

The final set of possible worlds is $\mathcal{W} = \{w_1, w_2, \ldots, w_{l+1}\}$, the probability distribution is $\mu(w_i)$, and $\Sigma_i \mu(w_i) = 1$. ii is a function on $\mathcal{A}_0$, and i is a function on $\mathcal{A}$.

We have the following theorem.

Theorem 5 *Given $(\mathcal{A}, Prob_*)$ where $\mathcal{A}$ is a set of axioms closed under $\wedge$ and $Prob_*$ is an assignment of lower bounds of probabilities on $\mathcal{A}$.*

Functions i and ii obtained from Algorithm B are an incidence function and a basic incidence assignment, respectively.

The corresponding GICT $< \mathcal{W}, \mu, P, \mathcal{A}, i >$ can produce $Prob_*$ on $\mathcal{A}$.

Proof.

Part I: ii and i are a basic incidence assignment and an incidence function respectively.

For any two formulae ϕ and ψ in $\mathcal{A}_0$ ($\phi \neq false$, $\psi \neq false$), we have:

$$\begin{aligned}&ii(\phi) \neq \emptyset,\ ii(\psi) \neq \emptyset\\ &ii(\phi) \cap ii(\psi) = \emptyset \ \ (when \ \ \phi \neq \psi),\\ &\cup_{\phi_j \in \mathcal{A}} ii(\phi_j) = \mathcal{W}\end{aligned}$$

Also, $ii(false) = \emptyset$, so ii is a basic incidence assignment. Therefore $i(\psi) = \cup_{\phi_j \models \psi} ii(\phi_j)$ is an incidence function on $\mathcal{A}$ based on Theorem 3.
The corresponding GICT is

$$< \mathcal{W}, \mu, P, \mathcal{A}, i > .$$

Part II: The lower bounds of probabilities of formulae $\mu(i_*(\phi))$ is the same as the original numerical assignment $Prob_*(\phi)$.

For any $\psi \in \mathcal{A}$, we calculate the lower bound of its probability, denoted by p_{i*} (in order to distinguish it from $Prob_*$), as follows:

$$\begin{aligned} p_{i*}(\psi) &= \mu(i_*(\psi)) \\ &= \mu(\cup_{\phi_j \in \mathcal{A}, \phi_j \models \psi} i(\phi_j)) \\ &= \mu(\cup_{\phi_j \in \mathcal{A}, \phi_j \models \psi} \cup_{\phi_{jl} \in \mathcal{A}, \phi_{jl} \models \phi_j} ii(\phi_{jl})) \\ &= \mu(\cup_{\phi_{jl} \in \mathcal{A}, \phi_{jl} \models \psi} ii(\phi_{jl})) \\ &= \mu(\cup_{\phi_{jl} \in \mathcal{A}, \phi_{jl} \models \psi, \phi_{jl} \neq \psi} ii(\phi_{jl})) + \mu(ii(\psi)) \\ &= \Sigma_{\phi_{jl} \in \mathcal{A}, \phi_{jl} \models \psi} \mu(ii(\phi_{jl})) + \mu(ii(\psi)) \\ &= \Sigma_{\phi_{jl} \in \mathcal{A}, \phi_{jl} \models \psi} Prob'_*(\phi_{jl}) + Prob'_*(\psi) \\ &= Prob_*(\psi). \end{aligned}$$

Therefore, this theory produces the same lower bounds of probabilities for those formulae in $\mathcal{A}$ as $Prob_*$.

QED

This algorithm is entirely based on the result that $ii(\phi) \cap ii(\psi) = \emptyset$.
In Algorithm B, for a formula ϕ, we keep deleting those portions of $Prob_*(\phi)$ which can be carried by formulae ϕ_j (where $\phi_j \models \phi$) until we obtain the last portion which must be carried by ϕ itself. This last portion will only be contributed by its basic incidence set.

4.2 Unique Output of the Algorithm

In this section we first re-consider Example 2.6 to demonstrate the importance of Algorithm B. Then we prove that the output of the algorithm is unique regardless of the order of axioms being selected. By uniqueness, we mean that *only one* basic incidence assignment can be mapped to a given numerical assignment.

From this basic incidence assignment, ii, we can generate as many incidence assignments as we want using Theorem 4. These generated incidence assignments are consistent with the given numerical assignment, and have the same basic incidence assignment.

We will also prove that for any consistent incidence assignment to the given numerical assignment, its basic incidence assignment is the same as the one obtained from Algorithm B (ii), if this incidence assignment shares the same set of possible worlds as the one created from Algorithm B.

Therefore, a basic incidence assignment derived from Algorithm B is enough to represent all the consistent incidence assignments, when these incidence assignments have the same set of possible worlds. [1]

Example 4.1

The input to Algorithm B is a set of axioms, $\mathcal{A} = \{a, b, c, a \wedge b, a \wedge c, b \wedge c, a \wedge b \wedge c, false, true\}$, with the lower bound of a probability distribution $Prob_*$ given as:

$$\begin{array}{ll} Prob_*(a) = 0.760, & Prob_*(b) = 0.640, \\ Prob_*(c) = 0.480, & Prob_*(a \wedge b) = 0.525, \\ Prob_*(a \wedge c) = 0.350, & Prob_*(b \wedge c) = 0.225, \\ Prob_*(a \wedge b \wedge c) = 0.165, & Prob_*(true) = 1, \\ Prob_*(false) = 0. & \end{array}$$

The set $\mathcal{A}$ is closed under the operator $\wedge$. Using Algorithm B, an incidence function is defined in the following steps:

Step 1. The set $\mathcal{A}_0$ is $\{a \wedge b \wedge c\}$ which contains the smallest formula in $\mathcal{A}$. This means that there is one possible world, $w_1 \in \mathcal{W}$, supporting formula $a \wedge b \wedge c$ and $\mu(w_1) = 0.165$.

We also have:

$$\begin{array}{l} Prob'_*(a \wedge b \wedge c) \leftarrow Prob_*(a \wedge b \wedge c) = 0.165, \\ \mu(w_1) \leftarrow 0.165, \\ ii(a \wedge b \wedge c) \leftarrow \{w_1\}, \\ \mathcal{A}' \leftarrow \mathcal{A} \setminus (\mathcal{A}_0 \cup \{false\}) \\ l \leftarrow 1 \end{array}$$

Step 2. $\mathcal{A}'$ still has more formulae than just tautologies.

Step 3. Choose formula $a \wedge b$ from $\mathcal{A}'$. Because $a \wedge b \wedge c$ is the only formula in $\mathcal{A}_0$, and it has the property that $a \wedge b \wedge c \models a \wedge b$, then we have

$$Prob'_*(a \wedge b) \leftarrow Prob_*(a \wedge b) - Prob'_*(a \wedge b \wedge c) = 0.525 - 0.165 = 0.36.$$

Since $Prob'_*(a \wedge b) > 0$, we have $\mathcal{W} \leftarrow \mathcal{W} \cup \{w_2\}$ and define:

[1] In Section 4.3, we will examine the situation when these incidence assignments do not share the same set of possible worlds.

$$ii(a \wedge b) \leftarrow \{w_2\},$$
$$\mu(w_2) \leftarrow Prob'_*(a \wedge b),$$
$$\mathcal{A}_0 \leftarrow \mathcal{A}_0 \cup \{a \wedge b\},$$
$$\mathcal{A}' \leftarrow \mathcal{A}' \setminus \{a \wedge b\},$$
$$l \leftarrow 2$$

Repeating this step for each of the remaining elements in $\mathcal{A}'$, we get:

$$\begin{array}{llll} \mathcal{W} \leftarrow \mathcal{W} \cup \{w_3\}, & ii(a \wedge c) \leftarrow \{w_3\}, & \mu(w_3) \leftarrow 0.185, & \mathcal{A}_0 \leftarrow \mathcal{A}_0 \cup \\ & & & \cup \{a \wedge c\}; \\ \mathcal{W} \leftarrow \mathcal{W} \cup \{w_4\}, & ii(b \wedge c) \leftarrow \{w_4\}, & \mu(w_4) \leftarrow 0.06, & \mathcal{A}_0 \leftarrow \mathcal{A}_0 \cup \\ & & & \cup \{b \wedge c\}; \\ \mathcal{W} \leftarrow \mathcal{W} \cup \{w_5\}, & ii(a) \leftarrow \{w_5\}, & \mu(w_5) \leftarrow 0.05, & \mathcal{A}_0 \leftarrow \mathcal{A}_0 \cup \{a\}; \\ \mathcal{W} \leftarrow \mathcal{W} \cup \{w_6\}, & ii(b) \leftarrow \{w_6\}, & \mu(w_6) \leftarrow 0.055, & \mathcal{A}_0 \leftarrow \mathcal{A}_0 \cup \{b\}; \\ \mathcal{W} \leftarrow \mathcal{W} \cup \{w_7\}, & ii(c) \leftarrow \{w_7\}, & \mu(w_7) \leftarrow 0.07, & \mathcal{A}_0 \leftarrow \mathcal{A}_0 \cup \{c\} \end{array}$$

$l = 7;$

Step 4. As

$$\begin{aligned} & \Sigma_j Prob'_*(\phi_j) \\ = \quad & 0.165 + 0.36 + 0.185 + 0.06 + 0.05 + 0.055 + 0.07 \\ = \quad & 0.945, \end{aligned}$$

then we have

$$\begin{aligned} \mathcal{W} & \leftarrow \mathcal{W} \cup \{w_8\}, \\ \mu(w_8) & \leftarrow 1 - \Sigma_j \mu(ii(\phi_j)) = 1 - \mu(\{w_1, \ldots, w_7\}) = 0.055, \\ \mathcal{A}_0 & \leftarrow \mathcal{A}_0 \cup \{true\}, \\ ii(true) & \leftarrow \{w_8\}. \end{aligned}$$

Step 5.

$$\begin{aligned} \mathcal{A}_0 & \leftarrow \mathcal{A}_0 \cup \{false\}, \\ ii(false) & \leftarrow \emptyset. \end{aligned}$$

The incidence function derived from $i(\phi) = \cup_{\phi_j \models \phi} ii(\phi_j)$ on the axiom set $\mathcal{A}$ is as shown below:

$$\begin{aligned} i(a \wedge b \wedge c) & = \{w_1\}, \\ i(a \wedge b) & = \{w_1, w_2\}, \\ i(a \wedge c) & = \{w_1, w_3\}, \\ i(b \wedge c) & = \{w_1, w_4\}, \end{aligned}$$

$$\begin{aligned} i(a) &= \{w_1, w_2, w_3, w_4\}, \\ i(b) &= \{w_1, w_2, w_4, w_6\}, \\ i(c) &= \{w_1, w_3, w_4, w_7\}, \\ i(true) &= \mathcal{W}, \\ i(false) &= \emptyset. \end{aligned}$$

We obtain $\mathcal{W} = \{w_1, \ldots, w_8\}$ with probability distribution μ on it. ii is a basic incidence assignment on $\mathcal{A}_0$.

◊

For any other formula ψ, if $\mu(ii(\psi)) = 0$, we explain this in two ways:

- there is no possible world making this formula true, or
- the probability of the subset of possible worlds making ψ true is 0.

In any case, it doesn't matter whether or not we add $ii(\psi)$ to the whole set of possible worlds. The corresponding GICT, $< \mathcal{W}, \mu, P, \mathcal{A}, i >$, can always produce the lower bound of a probability distribution $Prob_*$ on $\mathcal{A}$.

Algorithm B was implemented on a Sun Sparc 4 Workstation in Sicstus Prolog 2.1. The execution time for this example was 0.759 (seconds) in the order $\{a \wedge b \wedge c, a \wedge c, b \wedge c, a, b, a \wedge b, c, false, true\}$.

If the axioms are reordered as $\{a, b, c, b \wedge c, a \wedge c, a \wedge b, a \wedge b \wedge c, false, true\}$, there is little difference. The runtime for the latter case is 1.189 (seconds). The algorithm creates a set of possible worlds with 8 elements in both cases.

When we apply Algorithm B, there may be more than one formula satisfying the conditions in Step 2, but the order of choosing these formulae has no effects on the final result. In this example, after we choose $a \wedge b \wedge c$ in Step 1 and come to Step 2, it does not matter whether we choose $a \wedge b$ or $a \wedge c$ first. The final result remains the same.

This is proved in the following theorem:

Theorem 6 *Applying Algorithm B to $(\mathcal{A}, Prob_*)$ produces the same result regardless of the order of selecting formulae in Step 2.*

Proof.

Assume that $\mathcal{A}_0 = \{\phi_1, \ldots, \phi_k\}$ after Step 1, and that there are two formulae ψ_1, ψ_2 satisfying the condition specified in Step 2.

In Step 1, for every $\phi_j \in \mathcal{A}_0$, we have:

$$Prob'_*(\phi_j) = Prob_*(\phi_j),$$

$$\mu(w_j) = Prob'_*(\phi_j),$$

and

$$ii(\phi_j) = \{w_j\}.$$

Assume that we choose ψ_1 first in Step 3, we have

$$Prob'_*(\psi_1) = Prob_*(\psi_1) - \Sigma_{\phi_j \in \mathcal{A}_0, \psi_j \models \psi_1} Prob'_*(\phi_j).$$

Now, we choose ψ_2 and obtain:

$$\begin{aligned} Prob'_*(\psi_2) &= Prob_*(\psi_2) - \Sigma_{\phi_j \in \mathcal{A}_0 \cup \psi_1, \phi_j \models \psi_2} Prob'_*(\phi_j) \\ &\qquad \text{(because } \psi_1 \text{ is a smallest axiom in } \mathcal{A}' \text{ now)} \\ &= Prob_*(\psi_2) - \Sigma_{\phi_j \in \mathcal{A}_0, \phi_j \models \psi_2} Prob'_*(\phi_j) \text{ (because } \psi_1 \not\models \psi_2\text{)}, \end{aligned}$$

which indicates that adding ψ_1 into set $\mathcal{A}_0$ has no effects on the outcome of $Prob'_*(\psi_2)$.

In the same way we can prove that choosing ψ_2 first and then ψ_1 gives exactly the same $Prob'_*(\psi_1)$ and $Prob'_*(\psi_2)$ as above. That is, $Prob'_*$ on set $\mathcal{A}_0 \cup \{\psi_1, \psi_2\}$ remains the same no matter which formula in $\{\psi_1, \psi_2\}$ is chosen first.

Similarly, we can prove the theorem for any set of formulae $\{\psi_1, \ldots, \psi_n\}$ satisfying the condition in Step 3.

QED

Applying Algorithm B to $(\mathcal{A}, Prob_*)$ only produces one basic incidence assignment, as the order of choosing axioms has no effect on the final output. Through this basic incidence assignment, not only can an incidence function be defined on $\mathcal{A}$, but also a family of incidence assignments can be obtained through Theorem 4, which would share the same set of possible worlds, but with a different set of axioms.

On the other hand, any consistent incidence assignment, which shares the same set of possible worlds with the basic incidence assignment (ii) derived from Algorithm B, should have ii as its basic incidence assignment.

This is proved in the next theorem:

Theorem 7 *Assume that ii is the basic incidence assignment derived from Algorithm B on set $\mathcal{A}_0$, with the set of possible worlds as $\mathcal{W}$, for a given $(\mathcal{A}, Prob_*)$.*

Then, any consistent incidence assignment $< \mathcal{W}, \mu, P, \mathcal{A}'i >$ to $(\mathcal{A}, Prob_*)$, which takes $\mathcal{W}$ as the set of possible worlds, has ii as its basic incidence assignment.

Proof.

Assume that ii is the basic incidence assignment on set $\mathcal{A}_0$ from Algorithm B, and $\mathcal{W}$ is the corresponding set of possible worlds.

Then for each $\phi \in \mathcal{A}_0$ ($\mathcal{A}_0 \subseteq \mathcal{A}$), there exists a $w \in \mathcal{W}$, where $ii(\phi) = \{w\}$.

We further assume that the basic incidence assignment from a consistent incidence assignment $< \mathcal{W}, \mu, P, \mathcal{A}', i >$ is ii_1 on set $\mathcal{A}_1$, then for any $\psi \in \mathcal{A}_1$, there is a subset W_ψ of $\mathcal{W}$ so that $ii_1(\psi) = W_\psi$.

Now we need to prove that $\mathcal{A}_1$ and $\mathcal{A}_0$ are equivalent (the same set) and $ii_1(\phi) = ii(\phi), \forall \phi \in \mathcal{A}_0$.

Let ψ be a formula in $\mathcal{A}_1$ where $ii_1(\psi) = W_\psi$ and W_ψ contains two or more possible worlds. Assume that $W_\psi = \{w_1, w_2, \ldots\}$, then the smallest formula that both w_1 and w_2 support is ψ.

However, w_1 and w_2 are created to support different smallest formulae in the algorithm and they cannot be in the same basic incidence set of one formula. Therefore, there is one and only one possible world in W_ψ for every $\psi \in \mathcal{A}_1$.

For any w in $\mathcal{W}$, assume that $ii(\phi) = \{w\}$ and $ii_1(\psi) = \{w\}$, then ϕ and ψ are semantically equivalent to each other, as they are both taken as the smallest formula supported by w, and there is only one such smallest formula.

In this way, we consider every w in $\mathcal{W}$ one after another, then we can prove that $\mathcal{A}_1$ and $\mathcal{A}_0$ are isomorphic, that $\mathcal{A}_1 = \mathcal{A}_0$. For every $\phi \in \mathcal{A}_1$, $ii(\phi) = ii_1(\phi) = \{w_\phi\}$.

Therefore, the basic incidence assignment of $< \mathcal{W}, \mu, P, \mathcal{A}'i >$ is ii.

QED

There are two algorithms in [McLean, 1992] and [McLean *et al.*, 1995] for assigning incidences on axioms based on a probability assignment, one of which is the extension of a method given in [Bundy, 1986] in the original incidence calculus.

The common feature of the two algorithms is that a set of possible worlds has to be fixed first, such as a set of 100 elements, each of which with probability 1/100.

Given a set of axioms, both the algorithms try to divide possible worlds into groups and assign each group to an axiom. Therefore, not only the total number of axioms but also the interrelationship among these axioms affect the division procedure.

For instance, assume that there are only two distinct axioms a, b in $\mathcal{A}$, then the assignment procedure could be simply done by choosing two subsets of the set of possible worlds which can produce $Prob_*(a)$ and $Prob_*(b)$ respectively.

However, if $Prob_*(a \wedge b)$ is known as well in addition to $Prob_*(a)$ and $Prob_*(b)$, then not only two subsets of possible worlds W_1, W_2 are needed to match $Prob_*(a)$ and $Prob_*(b)$, but also another subset W_3 is required to match $Prob_*(a \wedge b)$ with the condition that $W_3 = W_1 \cap W_2$.

That is, the complexity of these two algorithms increases considerably along with the interrelationship among axioms. Because of this, the order of axioms also affects the efficiency of the algorithms.

In summary, there are three factors associated with the complexity of each algorithm in [McLean, 1992], [McLean *et al.*, 1995]: the total number of axioms, the relations among axioms, and the order of axioms in addition to the requirement of a fixed number of possible worlds.

However, in our algorithm, only one factor affects the complexity, that is, the number of axioms. Besides, the new algorithm does not require a set of possible worlds to be predefined.

Example 4.1 and Example 2.6 are about the same problem. Example 4.1 produces a set of possible worlds with 8 elements, while the two approaches in Example 2.6 use a predefined set of possible worlds containing 100 elements, each of which with 1/100 probability. Although there are only 8 axioms in $\mathcal{A}$, the algorithms in Example 2.6 take a long time to find a consistent assignment of incidences (with runtime 374.520 and 355.060 seconds respectively). Algorithm B needs only 1.189 (seconds) runtime to find a consistent assignment (all these experiments were carried out on a Sun Sparc 4 station in Sicstus Prolog).

This example illustrates that the relations among axioms could slow the algorithms down enormously, much worse than the size of set of axioms. In real world cases, axioms always have some interrelations.

4.3 Similarity of Separate Incidence Assignments

In the last section, we concluded that the basic incidence assignment obtained from Algorithm B subsumes all the incidence assignments which are consistent with the given numerical assignment, provided these incidence assignments have a common set of possible worlds, which is only relevant to the size of $\mathcal{A}$.

What would happen if there is an incidence assignment which is consistent with the numerical assignment, but does not use the common set of possible worlds generated from Algorithm B? Would this incidence assignment still be subsumed by the basic incidence assignment?

In this section, we will examine the relationships among different sets of possible worlds (including their probability distributions). In particular, we need to investigate the relationships between the probability space $(\mathcal{W}, 2^{\mathcal{W}}, \mu)$ obtained from Algorithm B and any other probability space from which $Prob_*$ on $\mathcal{A}$ can also be generated, provided there is a consistent i.

The research result given below shows that the probability space $(\mathcal{W}, 2^{\mathcal{W}}, \mu)$ obtained from Algorithm B subsumes all the possible probability spaces $(\mathcal{W}_j, 2_j^{\mathcal{W}}, \mu_j)$ from all consistent GICTs from any possible incidence assignment approach. In other words, Algorithm B does not omit any alternative incidence assignments implied by the numerical assignment.

In this section, we summarize the research carried out in [McBryan, 1996] which was later published in [Liu *et al.*, 1998]. We argue that one basic incidence assignment is the only possible outcome of Algorithm B. All the consistent incidence assignments from a numerical assignment can be generated from this unique basic incidence assignment no matter which set of possible worlds is involved.

Definition 4.1: *A regular GICT*

A GICT $< \mathcal{W}, \mu, P, \mathcal{A}, i >$ *is said to be regular iff*

$$\forall w \in \mathcal{W}, \mu(w) \neq 0.$$

That is, there are no possible worlds with zero probability. Any GICT can be converted to an equivalent regular theory by simply removing the excess possible worlds with zero probability.

For each w with $\mu(w) = 0$, we redefine:

$$\begin{aligned} &\mathcal{W}' = \mathcal{W} \setminus \{w\}, \\ &\forall w \in \mathcal{W}', \mu'(w) = \mu(w), \\ &\forall \phi \in \mathcal{A}, i'(\phi) = i(\phi) \setminus \{w\}. \end{aligned}$$

In the rest of this section, we assume that all GICTs have already been transformed in this way, if necessary, and when we refer to two GICTs as being equal, we mean subject to removal of these possible worlds.

Definition 4.2: *Similarity of possible worlds*

Given a GICT, $< \mathcal{W}, \mu, P, \mathcal{A}, i >$, *two possible worlds* $w_1, w_2 \in \mathcal{W}$ *are similar* $(w_1 \sim w_2)$ *iff*

$$\forall \phi \in \mathcal{A}, w_1 \in i(\phi) \Longleftrightarrow w_2 \in i(\phi).$$

Clearly, similarity is an equivalence relation. It is reflexive $(w \sim w)$, symmetric $(w_1 \sim w_2 \Rightarrow w_2 \sim w_1)$ and transitive $(w_1 \sim w_2 \wedge w_2 \sim w_3 \Rightarrow w_1 \sim w_3)$.

Therefore, we can talk about equivalence classes of similar possible worlds, and indeed in any GICT, $\mathcal{W}$ will be partitioned by the set of all such classes.

Definition 4.3: *Fundamental GICT*

A GICT, $< \mathcal{W}, \mu, P, \mathcal{A}, i >$*, is fundamental iff it has no distinct similar possible worlds, that is, iff*

$$\forall w_1, w_2 \in \mathcal{W}, w_1 \sim w_2 \Rightarrow w_1 = w_2.$$

Definition 4.4: *Direct subsumption of GICT*

A GICT, $< \mathcal{W}', \mu', P, \mathcal{A}, i' >$*, is directly subsumed by* $< \mathcal{W}, \mu, P, \mathcal{A}, i >$ *iff:*

- *For some* $w_1', w_2' \in \mathcal{W}'$ *with* $w_1' \sim w_2'$ *and some* $w_3 \notin \mathcal{W}'$*,* $\mathcal{W} = (\mathcal{W}' \setminus \{w_1', w_2'\}) \cup \{w_3\}$*,*
- $$\mu(w) = \begin{cases} \mu'(w) & : w \neq w_3 \\ \mu'(w_1') + \mu'(w_2') & : w = w_3 \end{cases}$$
- $$i(\phi) = \begin{cases} i'(\phi) & : w_1' \notin i'(\phi) \\ (i'(\phi) \setminus \{w_1', w_2'\}) \cup \{w_3\} & : w_1' \in i'(\phi) \end{cases}$$

In this definition, the fact that $w_1' \in i'(\phi)$ will imply that $w_2' \in i'(\phi)$, since $w_1' \sim w_2'$. So a subsumption replaces two possible worlds which occur in the same set of incidences with a single possible world whose probability is the sum of the probabilities of the previous two.

Definition 4.5: *Subsumption of GICTs*

A GICT $< \mathcal{W}_0, \mu_0, P, \mathcal{A}, i_0 >$ *is subsumed by* $< \mathcal{W}_n, \mu_n, P, \mathcal{A}, i_n >$ *iff there is a list of GICTs*

$$[< \mathcal{W}_0, \mu_0, P, \mathcal{A}, i_0 >, \ldots, < \mathcal{W}_n, \mu_n, P, \mathcal{A}, i_n >]$$

such that $\forall j = 1, \ldots, n$*,* $< \mathcal{W}_{j-1}, \mu_{j-1}, P, \mathcal{A}, i_{j-1} >$ *is directly subsumed by* $< \mathcal{W}_j, \mu_j, P, \mathcal{A}, i_j >$*.*

It is worth pointing out that the list can be singleton. That is, any GICT subsumes itself.

We have the following theorem:

Theorem 8 *A fundamental GICT can only be subsumed by itself.*

Proof.

Assume that a GICT subsumes a fundamental one, then there will be a chain of direct subsumptions between the two.

A direct subsumption acts on a pair of similar worlds in a GICT, but there can be no such pair in a fundamental theory and so the chain must have zero length.

Therefore, the fundamental GICT is subsumed by itself.

QED

Since probabilities of formulae (or lower bounds of probabilities, to be precise) are calculated through the incidence sets of these formulae, we can have the following statement:

If $< \mathcal{W}_1, \mu_1, P, \mathcal{A}, i_1 >$ subsumes $< \mathcal{W}_2, \mu_2, P, \mathcal{A}, i_2 >$, then

$$\forall \phi \in P, \mu_1(i_1(\phi)) = \mu_2(i_2(\phi))$$

where μ applied to a set is the sum of the results of applying μ to the individual members of that set.

And we have:

Theorem 9 *Subsumption preserves lower bounds of probability.*

Proof.

As a subsumption is made up purely of a series of direct subsumptions, it is sufficient to show that a direct subsumption preserves lower bounds of probability.

This, however, is not difficult, as $Prob_*(i(\phi)) = \mu(i(\phi))$.

Let $< \mathcal{W}, \mu, P, \mathcal{A}, i >$ directly subsume $< \mathcal{W}', \mu', P, \mathcal{A}, i' >$. Let w be the new possible world which is introduced in Definition 4.4.

Now, if $w \notin i'(\phi)$, then

$$Prob'_*(\phi) = \mu'(i'(\phi)) = \mu'(i(\phi)) = Prob_*(\phi).$$

If, on the other hand, $w \in i'(\phi)$, then:

$$\begin{aligned}
Prob'_*(\phi) &= \mu'(i'(\phi)) \\
&= \mu'((i(\phi) \setminus \{w_1, w_2\}) \cup \{w\}) \\
&= \mu'(i(\phi) \setminus \{w_1, w_2\}) + \mu'(w) \\
&= \mu(i(\phi) \setminus \{w_1, w_2\}) + \mu(w_1) + \mu(w_2) \\
&= \mu((i(\phi) \setminus \{w_1, w_2\}) \cup \{w_1\} \cup \{w_2\}) \\
&= \mu(i(\phi)) \\
&= Prob_*(\phi),
\end{aligned}$$

which proves the theorem.

QED

Note that because subsumption preserves lower bounds of probabilities in the above way, it also preserves similarity of possible worlds.

To illustrate this, we first define the occurence function.

Definition 4.6: *Occurrence function*

In a GICT, $< \mathcal{W}, \mu, P, \mathcal{A}, i >$, the occurrence (o) is a function mapping the possible worlds (in $\mathcal{W}$) to the subsets of the axioms (in $\wedge(\mathcal{A})$ which means that $\mathcal{A}$ is closed under $\wedge$) defined by

$$o(w) = \{\phi \in \mathcal{A} \mid w \in i(\phi)\}.$$

So, the occurrence of a possible world is the set of axioms to which this possible world supports. It is then trivial to show from Definition 4.2 that

$$w_1 \sim w_2 \iff o(w_1) = o(w_2).$$

Subsumption preserves occurrence (and hence similarity) since a possible world in a subsumption theory will either have existed in the original GICT with the same occurrence, or it will have replaced two similar possible worlds. In the later case, the occurrence of the new possible world will be (by Definitions 4.4 and 4.5) the same as that of the first two, and it will therefore be similar to any other possible world which was similar to the first two.

Definition 4.7: *Fundamental Subsumption*

A GICT has a fundamental subsumption iff it is subsumed by a fundamental GICT.

And we have:

Theorem 10 *Fundamental subsumption is unique up to renaming.*

Proof.

Let us assume that a GICT, $< \mathcal{W}, \mu, P, \mathcal{A}, i >$, is subsumed by two fundamental GICTs, $< \mathcal{W}_1, \mu_1, P, \mathcal{A}, i_1 >$ and $< \mathcal{W}_2, \mu_2, P, \mathcal{A}, i_2 >$.

We must provide an isomorphism between these to show that they are unique up to renaming. Clearly the sets of atomic propositions and axioms are the same in both cases.

Each direct subsumption acts on a pair of similar possible worlds, and replaces that pair with a single possible world. As stated earlier, subsumption preserves similarity of possible worlds. That is, for any equivalence class of $\sim$ and a direct subsumption acting on two possible worlds in it, the equivalence structure will be retained, with the substitution of the new possible world for the old two.

A fundamental subsumption continues this process until each equivalence class is reduced to a single element; if any of them has more than one, the GICT is not fundamental as similarities exist, and none can be reduced to zero.

If the number of equivalence classes in $\mathcal{W}$ is n, then the number of elements in both $\mathcal{W}_1$ and $\mathcal{W}_2$ will therefore also be n. We form an isomorphism between the two sets of possible worlds based on the original equivalence classes; an element $w_1 \in \mathcal{W}_1$ is mapped to $w_2 \in \mathcal{W}_2$ when w_1 and w_2 are subsumed from the same equivalence class in $\mathcal{W}$.

By Definition 4.4, μ_1 and μ_2 of each element will be the same of μ applied to all members of the corresponding equivalence set. As this original set is the same for both cases, μ_1 and μ_2 are isomorphic.

i_1 and i_2 are also isomorphic, this is clear as subsumption is incidence preserving.

QED

Morover, there holds:

Theorem 11 *Every GICT is subsumed by a fundamental GICT.*

Proof.

Let $< \mathcal{W}, \mu, P, \mathcal{A}, i >$ be a GICT. If it is fundamental, then it subsumes itself and we are done. If it is not fundamental, then there are two possible worlds, $w_1, w_2 \in \mathcal{W}$ such that $w_1 \sim w_2$.

We can therefore define a new GICT $< \mathcal{W}', \mu', P, \mathcal{A}, i' >$ which directly subsumes our original theory (by Definition 4.4).

Clearly, if $< \mathcal{W}', \mu', P, \mathcal{A}, i' >$ directly subsumes $< \mathcal{W}, \mu, P, \mathcal{A}, i >$ then $\mathcal{W}'$ has one less element than $\mathcal{W}$.

We can continue this process of direct subsumption, reducing the size of $\mathcal{W}$ at every step. As $\mathcal{W}$ is finite, and GICTs with less than two possible worlds are necessarily fundamental, we will eventually reach a fundamental GICT, which will subsume the original GICT by the definition of subsumption as a chain of direct subsumptions.

QED

So we have shown that every GICT is subsumed by one and only one fundamental GICT. We are now ready to define our equivalence relation on GICTs.

Definition 4.8: Similarity of GICTs

Two GICTs $< \mathcal{W}_1, \mu_1, P, \mathcal{A}, i_1 >$ and $< \mathcal{W}_2, \mu_2, P, \mathcal{A}, i_2 >$ are similar ($< \mathcal{W}_1, \mu_1, P, \mathcal{A}, i_1 > \sim < \mathcal{W}_2, \mu_2, P, \mathcal{A}, i_2 >$) iff they are subsumed by the same fundamental GICT.

It is clear that this is an equivalence relation, since every GICT is subsumed by itself, it is reflexive; the symmetricity is trivial; and the transitivity again follows trivially from the above results.

4.4 Fundamental Nature of Basic Incidence Assignments

On the basis of the previous discussion, it is necessary to prove that a GICT derived from applying the basic incidence assignment algorithm is fundamental. We must, before attempting to prove this, address the problem of possible worlds with zero probability. Algorithm B has ensured that no such possible worlds would have been created. So, the derived GICT is a regular one.

We have:

Theorem 12 *Any GICT derived from the basic incidence assignment obtained by applying Algorithm B to $(\mathcal{A}, Prob_*)$ ($\mathcal{A}$ is closed under $\wedge$) is fundamental, for that family of GICTs which have the same set of axioms.*

Proof.

Given a numerical assignment on a set of axioms $\mathcal{A}$, which is closed under $\wedge$, Algorithm B will produce a basic incidence assignment ii on $\mathcal{A}_0$ ($\mathcal{A}_0 \subseteq \mathcal{A}$). A number of incidence assignments can be derived from this basic incidence assignment on different sets of axioms (Theorems 3 and 4) where $\mathcal{A}$ is the smallest set among all of them.

We need to prove that every such derived GICT is fundamental to that family of GICTs which have the same set of axioms. To do this, we only need to take the GICT with $\mathcal{A}$ as the set of axioms as an example, as others can be proved in a similar way.

So, we need to prove that the GICT, $< \mathcal{W}, \mu, P, \mathcal{A}, i >$, derived from the basic incidence assignment ii on $\mathcal{A}_0$ ($\mathcal{A}_0 \subseteq \mathcal{A}$), is fundamental to all other GICTs in the form $< \mathcal{W}_j, \mu_j, P, \mathcal{A}, i_j >$ which share $\mathcal{A}$ but with different sets of possible worlds and probability distributions.

To prove this, we must show that no two distinct possible worlds in this GICT are similar – that is, for every pair of possible worlds w_1, w_2, there must be at least one $\phi \in \mathcal{A}$ such that either $w_1 \in i(\phi) \wedge w_2 \notin i(\phi)$ or $w_1 \notin i(\phi) \wedge w_2 \in i(\phi)$.

For simplicity, we will take w_1, w_2 to be ordered so that w_1 was created before w_2 in the algorithm. From the procedure of the algorithm, we know that there is a formula $\phi_1 \in \mathcal{A}$, that $ii(\phi_1) = \{w_1\}$, and likewise for w_2. Clearly, $w_1 \in i(\phi_1)$. We will show that $w_2 \notin i(\phi_1)$, which is sufficient, as w_1, w_2 are arbitrary.

In Algorithm B, we have

$$i(\phi_1) = \cup_{\phi_j \models \phi_1} ii(\phi_j).$$

So, $w_2 \in i(\phi_1)$ means that $ii(\phi_l) = \{w_2\}$ for a specific ϕ_l, because of the property of basic incidence assignment $ii(\phi_i) \cap ii(\phi_j) = \emptyset$. Therefore, $w_2 \in i(\phi_1)$ implies that $\phi_l \models \phi_1$.

However, if $\phi_l \models \phi_1$, then ϕ_1 should not have been selected before ϕ_l, which contradicts the choice of w_2. Therefore, this GICT is fundamental.

Similarly, as every other GICT derived from this basic incidence assignment has the same set of possible worlds, every such GICT must be fundamental.

QED

We have shown that every GICT is subsumed by a fundamental one, and that GICTs derived from Algorithm B are fundamental.

What we have to show next is that any GICT derived from any other incidence assignment approach is subsumed by a GICT from Algorithm B, given $(\mathcal{A}, Prob_*)$, where $\mathcal{A}$ is closed as usual.

Intuitively, the strict incidence set of a set of axioms is the set of possible worlds in which those axioms, and only those axioms, are true.

Definition 4.9: *Basic Probability Portion*

Given a GICT $< \mathcal{W}, \mu, P, A, i >$, *assume that* ii *is the basic incidence assignment derived from* i *on domain* $\mathcal{A}_0$.

Then, probability $\mu(ii(\phi))$ *for* $\phi \in \mathcal{A}_0$ *is called the basic probability portion carried by* ϕ, *denoted as* $bp(\phi)$. *For any* $\psi \in \mathcal{A} \setminus \mathcal{A}_0$, *we define* $bp(\psi) = 0$.

Clearly, following Definitions 3.7 and 4.9, we have

$$\sum_{\phi \in \mathcal{A}_0} bp(\phi) = 1.$$

The basic probability portion of an axiom is the portion of its lower bound of probability carried by that axiom, that cannot be obtained from other axioms.

There are the following corollaries:

Corollary 1 *Given a GICT, assume that the basic incidence assignment from it is* ii *on domain* $\mathcal{A}_0$, *then* w_1 *and* w_2 *are similar (Definition 4.2) if* $w_1, w_2 \in ii(\phi)$.

Proof.

The proof is straightforward due to Definitions 3.7 and 4.2, and Theorem 2.

QED

Corollary 2 *In a fundamental GICT, the basic probability portion of an axiom* ϕ *is either 0, when* $ii(\phi) = \emptyset$, *or* $\mu(w)$, *when* $ii(\phi) = \{w\}$.

Proof.

This follows directly from Definitions 3.7, 4.9 and 4.3.

QED

And now, we have:

Theorem 13 *A consistent lower bound of a probability distribution* $Prob_*$ *on a set of axioms* $\mathcal{A}$, *which is closed under conjunction, has one and only one fundamental consistent GICT, with* $\mathcal{A}$ *as the set of axioms.*

Proof.

From Theorems 3 and 4 we know that GICT, $< \mathcal{W}, \mu, P, \mathcal{A}, i >$, derived from Algorithm B has the smallest set of axioms which contains fewer elements than any other sets of axioms $\mathcal{B}_j$. We call this GICT the smallest theory in its family.

Assume that the corresponding GICTs of all consistent incidence assignments discovered from many other approaches are in set Q. Further assume that a subset of Q is Q_1 containing those GICTs which all have a common set of axioms, $\mathcal{A}$.

Then, we need to prove that this smallest theory is the only fundamental GICT subsuming all GICTs in Q_1. Similarly, we can prove that each GICT in $Q \setminus Q_1$ is subsumed by a GICT obtained through Theorem 4.

Part I:

Now, we will prove that the smallest GICT is the only fundamental GICT among all GICTs in Q_1.

Assume that we have a non-contradictory lower bound $Prob_*$ of a probability distribution on a finite set of n axioms $\mathcal{A}$, which is closed under conjunction.

The consistency of the lower bound ensures that a consistent incidence assignment is derivable from Algorithm B, which is fundamental. We must now show that it is unique.

If there are n axioms in $\mathcal{A}$, then there are at most $n+1$ possible worlds being created in Algorithm B. Each possible world w makes $ii(\phi) = \{w\}$ true (for a ϕ) and none for $false$. If the basic incidence assignment ii is defined on domain $\mathcal{A}_0$, then $bp(\phi) = \mu(w)$ for $\phi \in \mathcal{A}_0$ and $ii(\phi) = \{w\}$.

We now choose a GICT from Q_1 arbitrarily, say $< \mathcal{W}_1, \mu_1, P, \mathcal{A}, i_1 >$. There is a unique basic incidence assignment ii_1 derivable from i_1.

We assume that ii_1 is defined on domain $\mathcal{A}_1$. Then, based on Definition 4.9, we have $bp_1(\psi) = \mu_1(ii_1(\psi)) = \mu_1(W'), W' \subseteq \mathcal{W}_1$, for $\psi \in \mathcal{A}_1$ and $bp_1(\psi) = 0$ if $\psi \notin \mathcal{A}_1$.

However, as both bp and bp_1 define basic probability portions on axioms from the same consistent probability distribution, $bp(\phi)$ must be the same as $bp_1(\phi)$. Therefore, $\mathcal{A}_0$ and $\mathcal{A}_1$ specify the same set of axioms, for each of which the basic probability portion is not zero.

That is:

$$\mu(ii(\phi)) = \mu_1(ii_1(\phi)), \phi \in \mathcal{A}_0, \phi \in \mathcal{A}_1.$$

In other words, we have:

$$ii(\phi) = \{w\}, \mu(w) = \mu_1(ii_1(\phi)).$$

And:

(i) If for every $\phi \in \mathcal{A}_1$, $ii_1(\phi)$ has only one element w', then we can rename this possible world as w, where $ii(\phi) = \{w\}$.

So after renaming all the elements in $\mathcal{W}_1$, we have $\mathcal{W} = \mathcal{W}_1$. The smallest GICT is identical to $< \mathcal{W}_1, \mu_1, P, \mathcal{A}, i_1 >$.

(ii) If there exists an axiom ψ where $ii_1(\psi)$ has more than one element, say $ii_1(\psi) = \{w_1, w_2, \ldots, w_t\}$, then we have $w_1 \sim w_2 \sim \ldots \sim w_t$ (from Corollary 1).

Therefore, according to Definitions 4.4 and 4.5, the smallest GICT $< \mathcal{W}, \mu, P, \mathcal{A}, i >$ subsumes $< \mathcal{W}_1, \mu_1, P, \mathcal{A}, i_1 >$.

In summary, any GICT in Q_1, $< \mathcal{W}_1, \mu_1, P, \mathcal{A}, i_1 >$, is either identical to the smallest GICT, when the former is also a fundamental one, or subsumed by the smallest GICT. So, the smallest GICT is the unique fundamental GICT that subsumes all the GICTs in Q_1.

Part II:

Similarly, we can prove that any GICT from any incidence assignment approach is either identical to a GICT derived from Theorem 4 or subsumed by it.

Therefore, there is only one fundamental GICT for that family of GICTs which all have the same set of axioms, and the fundamental one is the one derived from the basic incidence assignment obtained through Algorithm B.

QED

4.5 Summary

In [Frisch and Haddawy, 1994] (page 117–118), the authors have stated that:

> "The major drawback of Bundy's approach is the lack of a general mechanism for assigning the initial incidences to a set of sentences, given their probabilities. The difficulty arises because each point[2] has a probability associated with it so that assigning incidences to sentences can result in specifying correlations among the sentences that were not specified in the original probabilities associated with the sentences. ... Assigning initial incidences from probabilities requires determining consistent truth assignments to the set of sentences. This problem is NP-complete for propositional logic and undecidable for first-order logic".

[2] Here, each point can be thought of as a possible world.

Two incidence assignment approaches in [McLean *et al.*, 1995] (briefly introduced in Chapter 2) are extremely slow, just because of the defficulty in determining consistent incidence assignments on a set of sentences, in particular, the correlations among sentences.

In this chapter, when we consider this problem again, we add an extra condition on the collection of those axioms (sentences), given their probabilities, i.e., we require the collection to be closed under operation $\wedge$. This is a crucial condition, as correlations among axioms are decidable under it. It would not be possible to design Algorithm B without this condition.

The most significant result of this chapter is that there is one and only one fundamental incidence assignment for any lower bound of a consistent probability distribution on a set of axioms which is conjunction, and this assignment is the one found by Algorithm B.

Any consistent incidence assignment produced by [McLean *et al.*, 1995] is therefore equivalent to this unique fundamental one. Furthermore, any consistent assignment produced by *any* method will be subsumed by this fundamental one. As there are an infinite number of such assignments (take any possible world and divide it into two, each with half the probability and repeat the procedure), Algorithm B is simpler than any other possible approach.

Chapter 5

Combining Multiple Pieces of Evidence

The third limitation of the original incidence calculus is related to the dealing with multiple pieces of evidence. We concluded that the original incidence calculus does not have the mechanism for doing this.

In fact, how to cope with multiple pieces of evidence has been a difficult task for any technique for reasoning under uncertainty.

In MYCIN's certainty factor model, there is a matching mechanism for information updating, when a new piece of evidence arrives [Shortliffe, 1976].

In probability theory, Bayesian's updating rule has been the tool for revising the beliefs when a new piece of evidence is available.

In DS theory, Dempster's combination rule is famous for its ability in combining two or more pieces of independent evidence [Shafer, 1976].

These combination mechanisms are all numerical based, and sometimes, produce results which are not so convincing or even contradictory (e.g., [Halpern and Fagin, 1992], [Zadeh, 1984], [Voorbraak, 1991]).

Researchers have been looking for alternative methods to deal with multiple pieces of evidence, [Halpern and Fagin, 1992] being an example. It is argued that combining evidence at the symbolic level would give better results than at the numerical level. However, there are few symbolic reasoning mechanisms which could support the development of an appropriate symbolic combination method.

When combining evidence in numerical reasoning approaches, a major difficulty is how to deal with overlapped information implied by multiple pieces of evidence. In other words, the difficulty is how to avoid using the same portion of information twice.

Furthermore, when numerical values are used to represent uncertainties, it is very difficult to tell how much the overlap is among two values. For example, given two probability values, 0.5 and 0.4 on a logical formula ϕ, it is almost impossible to judge how much overlap is implied in these two numerical values.

However, when symbolic methods are used to represent uncertainties, it is easy to tell the overlap by calculating the intersection of the two sets. By removing the intersection from one of the two sets, the overall information implied by the two sets can be manipulated. For instance, if two sets W_1 and W_2 are assigned to ϕ and $\mu(W_1) = 0.5$ and $\mu(W_2) = 0.4$, then the overlap of these two sets is $W_1 \cap W_2$.

Therefore, based on its symbolic and numerical features and the fact that inferences are performed at the symbolic level, incidence calculus is taken as an ideal mechanism to be experimented in searching for a better combination tool.

In this chapter, we aim at developing a combination mechanism at the symbolic level, which could overcome some limitations suffered by other purely numerical based combination methods, such as Dempster's combination rule.

5.1 Effects of New Information

Suppose we have already had a GICT, $< \mathcal{W}, \mu, P, \mathcal{A}_1, i_1 >$, for a given problem. If a new piece of information regarding this problem is known, then it may have one of the following effects:

Effect 1: The new source of information gives a new probability distribution on the set of possible worlds to *replace* the old probability distribution, then a new GICT is created to substitute for the old one and further inference is made solely upon the new GICT.

This can be seen through an example in [Fagin and Halpern, 1989a], [Fagin and Halpern, 1989b]. The example is stated as:

> Suppose that we have 100 agents, each holding a lottery ticket, numbered 00 to 99. Suppose that agent a_1 holds ticket number 17. Assume that the lottery is fair, so, a priori, the probability that a given agent will win is 1/100. We are then told that the first digit of the winning ticket is 1, the problem is to determine the probability that agent a_1 will win.

Considering this problem in generalized incidence calculus, we first form a GICT, $< \mathcal{W}, \mu_1, P, \mathcal{A}_1, i_1 >$, where $\mathcal{W} = \{00, \ldots, 99\}$, $\mu(w) = 1/100$, $P = \{a_1, \ldots, a_{100}\}$, $\mathcal{A}_1 = P$ and $i_1(a_i) = \{w\}$ when a_i's number is w. Here, a_i stands for the atomic proposition that a_i *will win.*

When we are told that *the first digit of the winning ticket is 1 later*, then the probability distribution on $\mathcal{W}$ is changed to $\mu_2(w) = 1/10$ when $w \in \{10, \ldots, 19\}$ and $\mu_2(w) = 0$ otherwise. Therefore, the new

GICT is $< \mathcal{W}, \mu_2, P, \mathcal{A}_1, i_1 >$. In this case, because the old probability distribution is replaced by the new one, the new GICT disables the effect of the old one. It is then easy to know that the probability that a_1 will win is 1/10.

It is interesting to notice that this new piece of evidence does not change the supporting relations between sets $\mathcal{L}(P)$ and $\mathcal{W}$. It only changes the function μ on $\mathcal{W}$.

Effect 2: The new source of evidence specifies a new incidence function between sets $\mathcal{L}(P)$ and $\mathcal{W}$ without changing the space of possible worlds. Then, a new GICT is formed. Both the new and old GICTs have an impact on $\mathcal{L}(P)$. So, it is necessary to consider how to obtain their joint impact.

For instance, assume that there is a piece of information which specifies the GICT

$$< \mathcal{W}, \mu, P, \mathcal{A}_1, i_1 >$$

where $\mathcal{W} = \{Mon, Tues, Wed, Thur, Fri, Sat, Sun\}$, $i_1(rainy) = \{Fri, Sat, Sun, Mon\}$, $i_1(windy) = \{Mon, Wed, Fri\}$, and $i_1(rainy \wedge windy) = \{Fri, Mon\}$.

If a new piece of information tells us that $i_2(rainy) = \{Fri, Sat, Sun\}$, $i_2(windy) = \{Wed, Fri\}$, and $i_2(rainy \wedge windy) = \{Fri\}$, then another GICT, $< \mathcal{W}, \mu, P, \mathcal{A}_2, i_2 >$, is formed which gives an alternative interrelation among the elements of the two sets (without changing the probability distribution on set $\mathcal{W}$).

We need to consider the joint impact of both the old and new pieces of information on formulae. That is, we must combine these two pieces of information. For a particular formula in $\mathcal{L}(P)$, if we have $i_1(\phi) = \mathcal{W}_1$ and $i_2(\phi) = \mathcal{W}_2$ from the two sources respectively, then the common impact of the two sources produces $i_1 \odot i_1(\phi) = \mathcal{W}_1 \cap \mathcal{W}_2$. Here, $\odot$ indicates that a kind of combination mechanism, to be defined later, is applied to i_1 and i_2.

More generally if $i_1(\phi) = \mathcal{W}_1$ and $i_2(\psi) = \mathcal{W}_2$, then $i_1 \odot i_2(\phi \wedge \psi) = \mathcal{W}_1 \cap \mathcal{W}_2$.

This is, the basic idea of creating a combination mechanism in incidence calculus which will be discussed in greater detail in the next section.

Effect 3 In contrast to the previous two effects, the new source of information leads to a new GICT based on a different set of possible worlds.

Like Effect 2, both the new and old GICTs have impact on $\mathcal{L}(P)$, so it is necessary to consider how to obtain their joint impact. If the old GICT is $< \mathcal{W}_1, \mu_1, P, \mathcal{A}_1, i_1 >$ and the new one is $< \mathcal{W}_2, \mu_2, P, \mathcal{A}_2, i_2 >$, then we form two sets of possible worlds. $\mathcal{W}_1$ with probability distribution

μ_1 and $\mathcal{W}_2$ with μ_2. However, in contrast to Effect 2, these two sets are not the same.

In summary, apart from some very simple cases shown in Effect 1, usually when new pieces of information are obtained it is necessary to combine them with the existing information. The corresponding combination mechanism is essential to play such a role in calculating the impact of all the information.

We now propose a combination mechanism in generalized incidence calculus for combining multiple pieces of evidence.

5.2 Combination Rule in Generalized Incidence Calculus

Definition 5.1: *Combination Rule*

Let $< \mathcal{W}, \mu, P, \mathcal{A}_1, i_1 >$, $< \mathcal{W}, \mu, P, \mathcal{A}_2, i_2 >$ *be two GICTs on the same set of possible worlds.*

Then, the joint impact of information carried by the them is represented by a quintuple:

$$< \mathcal{W} \setminus \mathcal{W}_0, \mu', P, \mathcal{A}, i >$$

where:

- $\mathcal{W}_0 = \bigcup\{i_1(\phi) \cap i_2(\psi) \mid (\phi \wedge \psi = false), \phi \in \mathcal{A}_1, \psi \in \mathcal{A}_2\}$.
- $\mathcal{A} = \{\varphi \mid \varphi = \phi \wedge \psi, \phi \in \mathcal{A}_1, \psi \in \mathcal{A}_2, \varphi \neq false\}$.
- $i(\varphi) = \bigcup\{i_1(\phi) \cap i_2(\psi) \mid (\phi \wedge \psi \models \varphi), \varphi \in \mathcal{A}, \phi \in \mathcal{A}_1, \psi \in \mathcal{A}_2, \phi \wedge \psi \neq false\}$.
- *For any* $w \in \mathcal{W} \setminus \mathcal{W}_0$, $\mu'(w) = \frac{\mu(w)}{1 - \Sigma_{w' \in \mathcal{W}_0} \mu(w')}$.
- $i(true) = \mathcal{W} \setminus \mathcal{W}_0$, $\quad i(false) = \emptyset$.

Here, $\phi \wedge \psi \neq false$ means $\phi \wedge \psi$ is not contradictory (invalid).

$\mathcal{W}_0$ is a subset of $\mathcal{W}$ reflecting the conflict of two pieces of information and the weight of conflict is $\Sigma_{w' \in \mathcal{W}_0} \mu(w')$.

If the weight of conflict is 1, these two pieces of information are completely contradictory with each other and they cannot be combined using the rule. $\mathcal{W}_0$ is taken off from $\mathcal{W}$ and the probability distribution on $\mathcal{W} \setminus \mathcal{W}_0$ is re-distributed.

When $\mathcal{A} \setminus \{true, false\} \neq \emptyset$, if $i(\phi) = \emptyset$ for $\forall \phi \in \mathcal{A} \setminus \{true, false\}$, then these two observations repel each other. In other words, only one of them can be held at a time.

We have the following theorem:

Theorem 14 *The quintuple obtained after applying Definition 5.1 to two GICTs is a GICT.*

Proof.

Let $< \mathcal{W}, \mu, P, \mathcal{A}_1, i_1 >$ and $< \mathcal{W}, \mu, P, \mathcal{A}_2, i_2 >$ be two GICTs.

To prove that $< \mathcal{W} \setminus \mathcal{W}_0, \mu', P, \mathcal{A}, i >$, the result of combining these two GICTs, is a GICT, we need to prove that:

1. $\mathcal{A}$ is closed under $\wedge$ (that is, when both φ_1 and φ_2 are in $\mathcal{A}$ and $\varphi_1 \wedge \varphi_2 \neq false$, $\varphi_1 \wedge \varphi_2$ is also in $\mathcal{A}$);
2. $i(\varphi_1 \wedge \varphi_2) = i(\varphi_1) \cap i(\varphi_2)$.

Part I:

Assume that φ_1, φ_2 are two distinct formulae ($\varphi_1 \neq \varphi_2$) in $\mathcal{A}$ and they are derived from formulae in $\mathcal{A}_1$ and $\mathcal{A}_2$ as $\varphi_1 = \phi_1 \wedge \psi_1$, and $\varphi_2 = \phi_2 \wedge \psi_2$, where $\phi_1, \phi_2 \in \mathcal{A}_1$ and $\psi_1, \psi_2 \subset \mathcal{A}_2$. We have $\varphi_1 \wedge \varphi_2 = (\phi_1 \wedge \phi_2) \wedge (\psi_1 \wedge \psi_2)$.

There are three situations, when $\varphi_1 \wedge \varphi_2$ could be in $\mathcal{A}$:

(i) When $\phi_1 \wedge \phi_2 = false$ or $\psi_1 \wedge \psi_2 = false$, then $\varphi_1 \wedge \varphi_2 = false$. $\varphi_1 \wedge \varphi_2$ does not need to be in $\mathcal{A}$.

(ii) When $i_1(\phi_1 \wedge \phi_2) = \emptyset$ or $i_2(\psi_1 \wedge \psi_2) = \emptyset$, then $i(\varphi_1 \wedge \varphi_2) = \emptyset$. $\varphi_1 \wedge \varphi_2$ does not need to be in $\mathcal{A}$.

(iii) Otherwise, $\phi_1 \wedge \phi_2 \in \mathcal{A}_1$ and $\psi_1 \wedge \psi_2 \in \mathcal{A}_2$, so $(\phi_1 \wedge \phi_2) \wedge (\psi_1 \wedge \psi_2) = (\phi_1 \wedge \psi_1) \wedge (\phi_2 \wedge \psi_2)$ is in the combined set $\mathcal{A}$. That is, $\varphi_1 \wedge \varphi_2$ is in $\mathcal{A}$.

Summarizing these three situations, we have that $\mathcal{A}$ is closed under $\wedge$.

Part II:

Next, we need to prove $i(\varphi_1 \wedge \varphi_2) = i(\varphi_1) \cap i(\varphi_2)$ when $\varphi_1, \varphi_2, \varphi_1 \wedge \varphi_2$ are all in $\mathcal{A}$.

Suppose that $i(\varphi_1 \wedge \varphi_2) \neq \emptyset$, for any $w \in \mathcal{W} \setminus \mathcal{W}_0$.

If $w \in i(\varphi_1 \wedge \varphi_2)$, then we have:

$$\begin{aligned}
& w \in i(\varphi_1 \wedge \varphi_2) \\
& \Rightarrow \exists \varphi_0 (\varphi_0 \models \varphi_1 \wedge \varphi_2) \wedge (w \in ii(\varphi_0)) \\
& \Rightarrow \exists \varphi_0 (\varphi_0 \models \varphi_1) \wedge (\varphi_0 \models \varphi_2) \wedge (w \in ii(\varphi_0)) \\
& \Rightarrow \exists \varphi_0 (i(\varphi_0) \subseteq i(\varphi_1)) \wedge (i(\varphi_0) \subseteq i(\varphi_2)) \wedge (w \in ii(\varphi_0)) \\
& \Rightarrow \exists \varphi_0 (i(\varphi_0) \subseteq i(\varphi_1) \cap i(\varphi_2)) \wedge (w \in ii(\varphi_0)) \\
& \Rightarrow w \in i(\varphi_1) \cap i(\varphi_2).
\end{aligned}$$

Table 5.1: The intuitive meaning of the combination rule

$r(i(r))$	$\phi_1(W_{11})$	$\dots$	$\phi_l(W_{1l})$		$\phi_n(W_{1n})$
$\psi_1(W_{21})$	$\phi_1\wedge$ $\wedge\psi_1(W_{11}\cap W_{21})$	$\dots$	$\dots$	$\dots$	$\dots$
$\vdots$					
$\psi_j(W_{2j})$	$\dots$	$\dots$	$\dots$	$\dots$	$\phi_l\wedge$ $\wedge\psi_j(W_{1l}\cap W_{2j})$
$\vdots$					
$\psi_m(W_{2m})$	$\dots$	$\dots$	$\dots$	$\dots$	$\dots$

The other way around, we have:

$$
\begin{aligned}
& w \in i(\varphi_1)\cap i(\varphi_2) \\
& \Rightarrow w \in i(\varphi_1)\wedge w \in i(\varphi_2) \\
& \Rightarrow (\exists\varphi_0(\varphi_0 \models \varphi_1)\wedge w \in ii(\varphi_0))\wedge(\exists\varphi_0'(\varphi_0' \models \varphi_2) w \in ii(\varphi_0')) \\
& \Rightarrow \exists\varphi_0(\varphi_0 \models \varphi_1)\wedge(\varphi_0 \models \varphi_2)\wedge(w \in ii(\varphi_0)) \\
& \qquad\qquad \text{(as } \varphi_0 \text{ and } \varphi_0' \text{ must be equivalent)} \\
& \Rightarrow \exists\varphi_0(\varphi_0 \models \varphi_1\wedge\varphi_2)\wedge(w \in ii(\varphi_0)) \\
& \Rightarrow \exists\varphi_0(i(\varphi_0)\subseteq i(\varphi_1\wedge\varphi_2))\wedge(w \in ii(\varphi_0)) \\
& \Rightarrow w \in i(\varphi_1\wedge\varphi_2).
\end{aligned}
$$

So,

$$i(\varphi_1\wedge\varphi_2) = i(\varphi_1)\cap i(\varphi_2).$$

As we have defined that $i(true) = \mathcal{W}\setminus\mathcal{W}_0$ and $i(false) = \emptyset$, function i is an incidence function. It is also easy to prove that $\Sigma_{w\in\mathcal{W}\setminus\mathcal{W}_0}\mu'(w) = 1$.

Therefore,
$<\mathcal{W}\setminus\mathcal{W}_0, \mu', P, \mathcal{A}, i>$ is a GICT.

QED

The explanation of this combination rule is that if the first observation says that $\mathcal{W}_{1i}$ makes formula ϕ true, and the second observation says that $\mathcal{W}_{2j}$ makes formula ψ true, then $\mathcal{W}_{1i}\cap\mathcal{W}_{2j}$ should make formula $(\phi\wedge\psi)$ true when we know that both X and Y hold.

The intuitive meaning of the rule is illustrated in Table 5.1. The straightforward application of the rule is shown in Example 5.1.

$r(i(r))$ in the table stands for a formula (r) and its incidence set ($i(r)$). The first row of Table 5.1 stands for the axioms and their incidence sets in the first GICT, such as W_{11} is the incidence set of axiom ϕ_1. The first column specifies those in the second theory.

The axioms in the combined theory are given in the lower right hand square of the table. Some axioms may be semantically equivalent to each

Table 5.2: Die numbers with corresponding urn numbers and colours of balls

die number	urn No.	ball colour
1, 2	urn I	Blue
3	urn II	Blue, Green
4, 5	urn III	Red
6	urn IV	Green, Red

Table 5.3: Two experts' knowledge

	die number	ball colour
X:	1,2,3	Blue, Green
	4,5,6	Red, Green
Y:	1,2	Blue
	3,4,5	Blue, Green, Red
	6	Green, Red

other and some may be logically entailed by others. So, the combined incidence set of a formula φ_{lj} is the union of all the incidence sets of those formulae appeared in the lower right hand square, each of which is either semantically equivalent to φ_{lj} or makes φ_{lj} true whenever it is true.

The underlying principle of calculating incidence sets of new axioms is

$$i(\varphi_{lj}) = i(\phi_l \wedge \psi_j) = \bigcup_{(\phi'_l \wedge \psi'_j) \models (\phi_l \wedge \psi_j)} i_1(\phi'_l) \cap i_2(\psi'_j).$$

where: ϕ_l and ψ_j are from two GICTs, respectively, and φ_{lj} is an axiom in the combined theory.

Example 5.1 (continuation of Example 3.5)

The example is:

> Assume that we have four urns full of different coloured balls. Which urn to draw a ball from is decided by the outcome of throwing a perfect die as shown in Table 5.2.
>
> However, two experts, X and Y, only know this information partially as shown in Table 5.3.
>
> Based on the two experts' knowledge, what is the best possible lower bound of the probability for statement "The colour of the drawn ball is blue"?

The two experts' knowledge is concerned with drawing a ball from urns, according to the same principle of drawing. So, the two pieces of information provided by the two experts specify two different but compatible mapping relations, i_1 and i_2, between two sets, the sets of die numbers and of colours. Therefore, it is possible to construct a GICT for each expert X and Y.

We have:

$$< \mathcal{W}, \mu, P, \mathcal{A}_1, i_1 >,$$

and

$$< \mathcal{W}, \mu, P, \mathcal{A}_2, i_2 >,$$

where $\mathcal{W} = \{1, 2, 3, 4, 5, 6\}$ is a set of possible worlds, each of which supports a particular outcome number of the thrown die (such as possible world 1 supports the outcome number 1) and $\mu(w_j) = 1/6$. P is a set of atomic propositions such as "the drawn ball has colour blue" (simply written as *Blue* in the following).

So, i_1 and i_2 are defined as:[1]

$$\begin{aligned} i_1(Blue \vee Green) &= \{1, 2, 3\}, \\ i_1(Red \vee Green) &= \{4, 5, 6\}, \\ i_1(true) &= \mathcal{W}, \\ i_1(false) &= \emptyset, \end{aligned}$$

on a set of axioms:

$$\mathcal{A}_1 = \{Blue \vee Green, Red \vee Green, true, false\},$$

and

$$\begin{aligned} i_2(Blue) &= \{1, 2\}, \\ i_2(Blue \vee Green \vee Red) &= \{1, 2, 3, 4, 5, 6\}, \\ i_2(Green \vee Red) &= \{6\}, \\ i_2(true) &= \mathcal{W}, \\ i_2(false) &= \emptyset, \end{aligned}$$

and on a set of axioms:

$$\mathcal{A}_2 = \{Blue, Blue \vee Green \vee Red, Red \vee Green, true, false\}.$$

These two GICTs are based on the same set of possible worlds, applying the combination rule in Definition 5.1 to them, we have

$$< \mathcal{W}, \mu, P, \mathcal{A}, i >,$$

[1] From Table 5.3, it is also possible to define two basic incidence assignments ii_1 and ii_2 and then construct i_1 and i_2. We leave this to the reader.

where:

$$\begin{aligned} i(Blue) &= \{1,2\}, \\ i(Blue \vee Green) &= \{1,2,3\}, \\ i(Green \vee Red) &= \{4,5,6\}, \\ i(Blue \vee Green \vee Red) &= \{1,2,3,4,5,6\}, \\ i(true) &= \mathcal{W}, \\ i(false) &= \emptyset, \end{aligned}$$

on set of axioms

$$\begin{aligned} \mathcal{A} = \; & \{Blue, Blue \vee Green, Green \vee Red, \\ & Blue \vee Green \vee Red, true, false\}. \end{aligned}$$

So, $i_*(Blue) = \{1,2\}$ and $Prob_*(Blue) = 2/6$. This result is consistent with what we can get from the precise information (Table 5.2).

However, for proposition "The colour of the drawn ball is Red", the precise information gives $Prob_*(Red) = 2/6$ while the combination result of the partial information from the two experts gives $Prob_*(Red) = 0$. This is because the experts' knowledge is not detailed enough to cover the precise information.

This example is adopted from [Chateauneuf, 1994] in which it can be solved in DS theory only under a series of mathematical descriptions (see [Chateauneuf, 1994] for details) because the two experts' knowledge is relevant. Dempster's combination rule is not applicable to it directly.

5.3 DS-independent Information

The central issue in applying the combination rule to two GICTs is that these two theories are based on the same set of possible worlds, even though sets of axioms and incidence functions might be different. The combination procedure applies the conjunctive operation to the axioms in the two theories to form new axioms, and the intersection operation to the incidence sets of axioms to obtain new incidence sets. In this way, a generalized incidence calculus is expected to be used to combine dependent evidence as well.

In general, there are three types of relations between two GICTs (provided by two pieces of evidence):

R1 The two sets of possible worlds in the two GICTs are the same. In this case, the combination rule is applied directly to combine them.

R2 The two sets of possible worlds (with their probability distributions) in the two GICTs are different and they are DS-independent[2]. In this

[2]According to [Voorbraak, 1991], two pieces of evidence are DS-independent, if their

case, it is possible to transform these two GICTs into new theories so that two new GICTs are based on the same set. Then, the combination rule is applicable to them. This is described in Proposition 5.1 below.

R3 The two sets of possible worlds in the two GICTs are different but not DS-independent. At the moment, we do not have a framework to deal with this type of relation in general. It has to be done individually. For a specific case in this type of relation, if it is possible to find a common set of possible worlds to replace the two existing spaces, then the combination rule is applicable. Otherwise, the generalized incidence calculus cannot cope with the case.

As cases in type R2 can be transformed into cases in type R1, then R2 is regarded as a special form of R1.

Definition 5.2: *DS-independence [restated from [Voorbraak, 1991]]*

Two probability spaces, (X_1, χ_1, μ_1) and (X_2, χ_2, μ_2), are said to be DS-independent, if they satisfy the conditions:

$$\begin{aligned} \mu_\chi(C_i \mid D_j) &= \mu_1(C_i), \\ \mu_\chi(D_j \mid C_i) &= \mu_2(D_j), \end{aligned}$$

for all subsets C_i and D_j, where: $\chi_1' = \{C_1, \ldots, C_n\}$ and $\chi_2' = \{D_1, \ldots, D_m\}$ are bases for χ_1 and χ_2, respectively. μ_χ is the (a priori) probability measure on χ which is the σ-algebra of the joint space X.

Spaces X_1 and X_2 are constructed from the joint space X. If the joint space is $X_1 \otimes X_2$, then $\mu_\chi(C_i)$ is an abbreviation for $\mu_\chi(\{(C_i, D_j) \mid 1 \leq j \leq m\})$.

It is easy to see that if two probability spaces are DS-independent, then they must be probabilistically independent.

If two probability spaces are DS-independent, then their common probability space is their set product, that is:

$$(X, \chi, \mu) = (X_1 \otimes X_2, \chi_1 \otimes \chi_2, \mu_1 \otimes \mu_2).$$

From now on, two GICTs are said to be DS-independent if their sets of possible worlds (with their probability distributions) are DS-independent, i.e., two probability spaces

$$(\mathcal{W}_1, 2^{\mathcal{W}_1}, \mu_1) \qquad (\mathcal{W}_2, 2^{\mathcal{W}_2}, \mu_2)$$

are DS-independent.

---

sources were not obtained from a common source. Dempster's combination rule is only applicable under this condition.

We give the definition of DS-independence in Definition 5.2 (see [Voorbraak, 1991] for the comprehensive discussion and explanation

Definition 5.3: *Transformation of GICTs*

Given two DS-independent GICTs,

$$< \mathcal{W}_1, \mu_1, P, \mathcal{A}_1, i_1 >$$

and

$$< \mathcal{W}_2, \mu_2, P, \mathcal{A}_2, i_2 >,$$

two new GICTs can be derived from them as

$$< \mathcal{W}_3, \mu_3, P, \mathcal{A}_1, i'_1 >,$$

and

$$< \mathcal{W}_3, \mu_3, P, \mathcal{A}_2, i'_2 >,$$

where:

- $\mathcal{W}_0 = \bigcup_{\phi \wedge \psi \models false} i_1(\phi) \otimes i_2(\psi), \mathcal{W}_3 = \mathcal{W}_1 \otimes \mathcal{W}_2 \setminus \mathcal{W}_0,$
- $i'_1(\phi) = (i_1(\phi) \otimes \mathcal{W}_2) \setminus \mathcal{W}_0$ *(where* $\phi \in \mathcal{A}_1$*),*
- $i'_2(\psi) = (\mathcal{W}_1 \otimes i_2(\psi)) \setminus \mathcal{W}_0$ *(where* $\psi \in \mathcal{A}_2$*).*

The new probability distribution on $\mathcal{W}_3$ *is:*

$$\mu_3(< w_{1i}, w_{2j} >) = \frac{\mu_1(w_{1i})\mu_2(w_{2j})}{1 - \Sigma_{< w_{1t}, w_{2m} > \in \mathcal{W}_0} \mu_1(w_{1t})\mu_2(w_{2m})}. \tag{5.1}$$

where: $\Sigma_{< w_{1t}, w_{2m} > \in \mathcal{W}_0} \mu_1(w_{1t})\mu_2(w_{2m})$ is the weight of the conflict between two theories. If the weight of conflict is 1 then these two pieces of information are completely in conflict with each other and they cannot be combined.

Proposition 5.1:

The combination rule is applicable to two DS-independent GICTs. The result is also a GICT.

The proof of this proposition can be easily obtained from Definition 5.3.

An element $< w_{1i}, w_{2j} >$ in $\mathcal{W}_1 \otimes \mathcal{W}_2 \setminus \mathcal{W}_0$ tells us that possible worlds w_{1i} and w_{2j} may both support a formula at the same time.

An element $< w_{2j}, w_{1i} >$ in $\mathcal{W}_2 \otimes \mathcal{W}_1$ implies the same meaning as $< w_{1i}, w_{2j} >$. Therefore, we treat $\mathcal{W}_1 \otimes \mathcal{W}_2$ and $\mathcal{W}_2 \otimes \mathcal{W}_1$ as the same set.

Hence, the combination rule is both commutative and associative because the result of combining several incidence calculus theories is unique irrespective of the order in which they are combined.

5.4 Examples

In this section we analyze some examples. The first two examples are simplified from [Shafer, 1986]. These are examples based on the same scenario but with different assumptions.

The first example assumes that the two pieces of evidence are DS-independent, while the second does not. In both cases the combination rule in extended incidence calculus is applicable to obtain correct results.

The third example is from [Pearl, 1988] which illustrates the situation that falls into relation type R3 among different GICTs. This example cannot be dealt with using the combination rule we proposed.

Example 5.2

This example can be stated as:

> Assume that there are two sensors, A and B, to observe a road condition. Sensor A is in perfect working condition, but sometimes gives wrong results. Past experience shows that A is accurate 80% of the time. Sensor B always gives accurate results when it is in working condition and there is a 99% chance that B is working properly. Sensor A shows that the road outside is frozen and B shows not. What is the true result given that A and B are independent?

Based on this story, several sets can be constructed. First of all, a set of atomic propositions is necessary with at least two propositions $P = \{frozen, not\_frozen\}$.

Then, set $\mathcal{W}_1$ is formed to show the accuracy of sensor A:

$$\mathcal{W}_1 = \{accurate, not\_accurate\},$$

with the probability distribution

$$\mu_1(accurate) = 0.8, \qquad \mu_1(not\_accurate) = 0.2$$

where *accurate* and *not_accurate* stand for "Sensor A is accurate" and "Sensor A is not accurate", respectively.

Finally, set $\mathcal{W}_2$ is also required regarding whether B is working properly:

$$\mathcal{W}_2 = \{working, not\_working\},$$

with the probability distribution

$$\mu_2(working) = 0.99, \qquad \mu_2(not\_working) = 0.01$$

where *working* and *not_working* stand for "Sensor B is working" and "Sensor B is not working", respectively.

Set $\mathcal{W}_1$ can be taken as a set of possible worlds interpreting the situations where "the road is frozen" and "the road may or may not be frozen" are true respectively. Similarly, set $\mathcal{W}_2$ can also be treated as a set of possible worlds, explaining the situations where statements "the road may or may not be frozen" and 'the road is frozen" are true, respectively.

Therefore, it is possible to construct two GICTs from the story:

$$< \mathcal{W}_1, \mu_1, P, \mathcal{A}_1, i_1 >,$$

and

$$< \mathcal{W}_2, \mu_2, P, \mathcal{A}_2, i_2 >,$$

where:

$\mathcal{A}_1 = \{frozen, true, false\}$,
$i_1(frozen) = \{accurate\},\ i_1(true) = \mathcal{W}_1,\ i_1(false) = \emptyset$,
$\mathcal{A}_2 = \{not_frozen, true, false\}$,
$i_2(not\_frozen) = \{working\},\ i_2(true) = \mathcal{W}_2,\ i_2(false) = \emptyset$.

Because of the DS-independence of the two sources, Proposition 5.1 in Section 5.3 can be used to combine these two theories and the third generalized incidence calculus theory $< \mathcal{W}, \mu, P, \mathcal{A}, i >$ can be obtained, where:

$$\mathcal{W}_0 = \{< accurate, working >\},$$

$$\mathcal{W} = \mathcal{W}_1 \otimes \mathcal{W}_2 \setminus \mathcal{W}_0,$$

$$\mu(< w_{1l}, w_{2j} >) = \frac{\mu_1(w_{1l}) \times \mu_2(w_{2j})}{1 - \Sigma_{< w'_{1l}, w'_{2j} > \in \mathcal{W}_0} \mu_1(w'_{1l}) \times \mu_2(w'_{2j})},$$

$$\mathcal{A} = \{frozen, not\_frozen, true, false\},$$

$$i(false) = \emptyset.$$

And, we have:

$$\begin{aligned} i(frozen) &= i(frozen \wedge true) = i_1(frozen) \otimes i_2(true) \setminus \mathcal{W}_0 \\ &= \{< accurate, not\_working >\}, \end{aligned}$$

$$\begin{aligned} i(not\_frozen) &= i(true \wedge not\_frozen) = i_1(true) \otimes i_2(not\_frozen) \setminus \mathcal{W}_0 \\ &= \{< not\_accurate, working >\}, \end{aligned}$$

$$\begin{aligned} i(true) &= (i_1(frozen) \otimes i_2(truc) \cup i_1(true) \otimes i_2(not\_frozen) \cup \\ &\quad \cup i_1(true) \otimes i_2(true)) \setminus \mathcal{W}_0 \\ &= \{< accurate, not\_working >, < not\_accurate, working >, \\ &\quad < not\_accurate, not\_working >\}, \end{aligned}$$

Table 5.4: Combination of two DS-independent evidence

We have ignored axiom "false" in both GICTs.

ϕ $i(\phi)$	$frozen$ $\{accurate\}$	$true$ $\mathcal{W}_1$
not_frozen $\{working\}$	$false$ $\{accurate\} \otimes \{working\}$	not_frozen $\mathcal{W}_1 \otimes \{working\}$
$true$ $\mathcal{W}_2$	$frozen$ $\{accurate\} \otimes \mathcal{W}_2$	$true$ $\mathcal{W}_1 \otimes \mathcal{W}_2$

Therefore, we have:

$$Prob_*(frozen) = \mu(i(frozen)) = 0.04$$

,

$$Prob_*(not\_frozen) = 0.95$$

which are the degrees of our belief in two propositions: "It is frozen" and "It is not frozen". Table 5.4 below shows the combination procedure of two GICTs in Example 5.2.

This example is classified into type R2 in Section 5.3,

◇

Example 5.3

In Example 5.2, we assume that sensors A and B are independent. However, if the accuracy of sensor A is affected by sensor B in the way that A is with 90% chance being *not_accurate* when B is not working, then the two sources cannot be regarded as independent to each other anymore. Under this assumption, what can we infer?

In this case, we still have three sets, P, $\mathcal{W}_1$ and $\mathcal{W}_2$, as well as the supporting relations between $\mathcal{W}_1$, $\mathcal{W}_2$ and P. However, we cannot deal with this case as we did in Example 5.2 because of the dependence between the elements in the sets $\mathcal{W}_1$ and $\mathcal{W}_2$.

We will have to find a common set out of $\mathcal{W}_1$ and $\mathcal{W}_2$ first and then establish the supporting relations between this common set and P. In this specific case, the common set is $\mathcal{W} = \mathcal{W}_1 \otimes \mathcal{W}_2$ but the probability distribution on it is no longer $\mu_1 \otimes \mu_2$.

The real probability distribution was given in [Shafer, 1986] as follows after having taken into account the fact the chance that A is *not_accurate* equals 90% when B is not working:

$$\mu(< accurate, working >) = 0.799,$$
$$\mu(< accurate, not\_working >) = 0.001,$$
$$\mu(< not\_accurate, working >) = 0.191,$$
$$\mu(< not\_accurate, not\_working >) = 0.009.$$

From the supporting relations between $\mathcal{W}_1, \mathcal{W}_2$ and P, two incidence functions are decidable from $\mathcal{W}$ to P as:

$$i_1(frozen) = \{< accurate, working >, < accurate, not\_working >\},$$

$$i_1(true) = \mathcal{W}, \ \ i_1(false) = \emptyset,$$

$$\mathcal{A}_1 = \{frozen, true, false\},$$

and

$$i_2(not\_frozen) = \{< accurate, working >, < not\_accurate, working >\},$$

$$i_2(true) = \mathcal{W}, \ \ i_2(false) = \emptyset,$$

$$\mathcal{A}_2 = \{not\_frozen, true, false\}.$$

So, the two corresponding GICTs are:

$$< \mathcal{W}, \mu, P, \mathcal{A}_1, i_1 >,$$

and

$$< \mathcal{W}, \mu, P, \mathcal{A}_2, i_2 > .$$

Combining them using our combination rule, we get the third GICT

$$< \mathcal{W}', \mu', P, \mathcal{A}, i >$$

where: $\mathcal{W}_0 = \{< accurate, working >\}$, $\mathcal{W}' = \mathcal{W} \setminus \mathcal{W}_0$, and μ' as shown below:

$$\mu'(< accurate, working >) = 0.0,$$
$$\mu'(< accurate, not\_working >) = 0.005,$$
$$\mu'(< not\_accurate, working >) = 0.950,$$
$$\mu'(< not\_accurate, not\_working >) = 0.45.$$

From this combined GICT, we obtain the degrees of our belief in *frozen* and *not_frozen* as 0.005 and 0.95, respectively.

This example is in relation type R3 in Section 5.3. Because we are able to construct the joint set and the probability distribution on it, our combination rule can be applied to solve it.

Example 5.4

We now use an example adopted from ([Pearl, 1988] pp.58) to show the situation in which two GICTs are based on different sets of possible worlds but these two sets are not DS-independent. In this case, the combination rule we proposed cannot deal with it properly.

The example is as follows:

> There are three prisoners, A, B and C, who have been tried for murder, and verdicts for them will be read tomorrow. They know only that one of them will be declared guilty and the other two will be set free. The identity of the condemned prisoner is revealed to the very reliable prison guard, but not to the prisoners themselves. In the middle of the night, Prisoner A calls the guard over and makes the following request: 'Please give this letter to one of my friends – to one who is to be released. You and I know that at least one of them will be freed'. Later, Prisoner A calls the guard again and asks who received the letter. The guard answers, 'I gave the letter to Prisoner B, he will be released tomorrow'. After this Prisoner A feels that his chance to be guilty has been increased from 1/3 to 1/2. What did he do wrong?

Assume that I_B stands for the proposition 'Prisoner B will be declared innocent' and G_A stands for the proposition 'Prisoner A will be declared guilty'. The task is to compute the probability of G_A given all the information obtained from the guard.

By solving this problem by formal probability theory, Pearl [Pearl, 1988] gets:

$$Pr(G_A \mid I_B) = \frac{Pr(I_B \mid G_A)Pr(G_A)}{Pr(I_B)} = \frac{Pr(G_A)}{Pr(I_B)} = \frac{1/3}{2/3} = 1/2 \qquad (5.2)$$

where: $Pr(I_B \mid G_A) = 1$ since $G_A \models I_B$, and $Pr(G_A) = Pr(G_B) = Pr(G_C) = 1/3$ from the prior probability distribution.

Pearl [Pearl, 1988] argues that this is a wrong result and the wrong result arises from omitting the *full context* in which the answer was obtained by Prisoner A. He further explains that 'By context we mean the entire range of answers one could possibly obtain, not just the answer actually obtained'. Therefore, Pearl introduces another proposition I'_B, stands for 'The guard said that B will be declared innocent', and he gives that:

$$Pr(G_A \mid I'_B) = \frac{Pr(I'_B \mid G_A)Pr(G_A)}{Pr(I'_B)} = \frac{1/2.1/3}{1/2} = 1/3 \tag{5.3}$$

which he believes is the correct result.

Using incidence calculus to solve this problem, we let $P = \{G_A, G_B, G_C\}$ and G_A stand for the proposition 'Prisoner A is guilty'.

Then it is possible to form a set of possible worlds $\mathcal{W}_1 = \{w_1, w_2, w_3\}$ with $\mu_1(w_j) = 1/3$ from the prior probability distribution, where possible worlds w_1, w_2 and w_3 support prisoners A, B and C are guilty, respectively.

From this information, a GICT is formed as:

$$< \mathcal{W}_1, \mu_1, P, \mathcal{A}, i_1 >$$

with $\mathcal{A} = \{G_A, G_B, G_C, truth, false\}$, and $i_1(G_A) = \{w_1\}$, $i_1(G_B) = \{w_2\}$ and $i_1(G_C) = \{w_3\}$.

After the guard passed the letter to a prisoner, it is possible to form another set of possible worlds $\mathcal{W}_2 = \{L_B, L_C\}$ where L_B means that Prisoner B received the letter; $\mu_2(L_B) = \mu_2(L_C) = 1/2$.

So, the second GICT is constructed as

$$< \mathcal{W}_2, \mu_2, P, \mathcal{A}_2, i_2 >$$

where $i_2(G_A \vee G_C) = \{L_B\}$, $i_2(G_A \vee G_B) = \{L_C\}$ and $\mathcal{A}_2 = \{G_A \vee G_C, G_A \vee G_B, truth, false\}$.

These two theories are based on different sets of possible worlds and they are not DS-independent. If we attempt to solve this example as if the two GICTs were DS-independent using Proposition 5.1, then we could have only obtained the result shown in equation (5.2).

However, whether it is possible to construct different GICTs in order to reflect the *full context* of answers (the meaning of I'_B not I_B), remains to be solved.

5.5 Summary

Intelligent systems (such as in planning, decision making, robotics, etc.) often require an ability to derive a consistent set of conclusions from a number of sources, which may be potentially conflicting. This process is known as information combination or fusion – a process to form a single combined view or perspective.

Available approaches include Bayesian updating in probability theory and bayesian belief networks, Dempster's combination rule in DS theory, etc. (see *Readings in Uncertainty Reasoning* [Shafer and Pearl, 1990] or review articles exemplified by, e.g., [Stephanou and Sage, 1986]) for details). The corresponding combination mechanisms in MYCIN [Shortliffe, 1976] and in PROSPECTOR [Duda *et al.*, 1976], can be formalized in a unified framework using algebraic approach and are proved to be isomorphic to each other [Spiegelhalter, 1986].

Bayesian updating is the safest approach when pieces of evidence are representable in probabilities. Dempster's combination rule is the most popular one when ignorance exists, or when there are difficulties in assigning probabilities. A delicate relationships between probability theory and DS theory is thoroughly discussed in [Halpern and Fagin, 1992]. In a situation when probability theory is not suitable, and when pieces of evidence are inter-related, neither of the above two methods is applicable.

In this chapter, we have consider combination problem from a different perspective. An attempt has been made in order to eliminate the overlapped information provided by two non-independent sources. Because generalized incidence calculus performs its inference symbolically, a combination rule was proposed at this level based on its unique structure. It is shown that the new combination rule is able to combine independent and some dependent pieces of evidence (we shall resume this topic again in Chapter 7). The creation of this rule has also broadened the application of generalized incidence calculus in practice.

Chapter 6

The Dempster–Shafer Theory of Evidence

The Dempster-Shafer theory of evidence (here, DS theory, for brevity), sometimes called evidential reasoning (cf. Lowrance et al. [Lowrance *et al.*, 1981]) or belief function theory, is a mechanism formalised by Shafer ([Shafer, 1976]) for representing and reasoning with uncertain, imprecise and incomplete information. It is based on Dempster's original work ([Dempster, 1967]) on the modelling of uncertainty in terms of upper and lower probabilities that are induced by a multivalued mapping rather than as a single probability value.

The transferable belief model [Smets and Kennes, 1994] was developed based on this theory.

DS theory reduces to standard Bayesian reasoning when an agent's knowledge is accurate but it is more flexible in representing and dealing with ignorance and uncertainty [Barnett, 1981].

DS theory has been popular since the early 1980s when AI researchers were searching for different mechanisms to cope with those situations where Bayesian probability was powerless. Its relationships with related theories have been intensively discussed in [Yager, Fadrizzi, Kacprzyk, 1994].

There are two main reasons why DS theory has attracted a lot of attention. It has an ability to model information flexibly, without requiring a probability (or a prior) to be assigned to each element in a set, and it provides a convenient and simple mechanism (Dempster's combination rule) for combining two or more pieces of evidence under certain conditions. The former allows a user to describe ignorance because of lack of information, and the latter allows an agent to narrow the space of possible answers as more evidence is accumulated.

Even though DS theory has been widely used, it was found that Dempster's combination rule gives counterintuitive results in many cases. The condition under which the rule is used is crucial to the successful application of the theory but the condition was not fully defined when Shafer gave the rule in the first instance [Shafer, 1976].

Many discussions and criticisms of Dempster's rule have appeared in the literature. The interested reader can find here details in, e.g., [Zadeh, 1986], [Dubois and Prade 1990c], [Dubois and Prade 1992b], [Kruse *et al.*, 1992], [Fagin and Halpern, 1990], [Voorbraak, 1991], [Lingras and Wong, 1990], [Nguyen and Smets, 1993], [Hunter,1987], [Pearl, 1988], [Black,1987].

A mathematical description of the condition of applying Dempster's combination rule is formalised in [Voorbraak, 1991] in which two pieces of evidence can be combined when they are DS-independent (see Definition 5.2).

In this chapter, we first introduce basic elements of DS theory, then study the probabilistic basis of mass functions defined by Shafer [Shafer, 1976] to see how a mass function can be derived from a probability distribution through a multivalued mapping [Dempster, 1967].

The discussion shows that DS theory is closely related to probability theory and provides a convenient way of describing the condition of using Dempster's combination rule.

Through examining Dempster's original idea on evidence combination in detail, we demonstrate that – in fact – Dempster suggested two alternative approaches for combining evidence. One approach requires the construction of a common source covering several original sources, and propagates the numerical values of uncertainty on the common source to the target set. Another approach encourages individual propagation from several original sources to the target space, and then combines propagation results at the target level. Dempster's combination rule in DS theory is a simplified form of Dempster's second approach.

We argue that a simplified form of combination (i.e. Dempster's combination rule) does not carry enough information to allow an agent to judge whether two pieces of evidence can be combined.

In fact, under Dempster's original idea, some counterintuitive examples of using Dempster's combination rule are not counterintuitive at all, because these examples cannot and should have not been dealt with in DS theory in an appropriate way.

Apart from its confusing conditions of combination, DS theory was also criticized for its large computational requirements, with the number of combinations being exponential in the size of a problem domain.

Several algorithms have been implemented to improve the computational complexity. The earliest attempt was given by Barnett [Barnett, 1981] in which the combination only involves a specific form of evidence. Since then, work has been done on designing more general mechanisms for efficient combination.

In this chapter, several typical efficient combination mechanisms are introduced. Potential efficiency gains through the use of parallel techniques are also highlighted.

In its application to expert systems, DS theory was found incapable of representing heuristic knowledge, such as, if a is true, then b is true with probability 0.7. Research has been done to extend the theory to cope with heuristic knowledge such as, e.g., [Yen, 1989], [Guan and Bell, 1991], [Liu *et al.*, 1994].

We will briefly review the probabilistic mapping in [Yen, 1989] and the evidential mapping in [Liu *et al.*, 1994] to see how to tackle this problem.

Discussions on other limitations of DS theory are also mentioned briefly.

6.1 Basic Concepts in the Dempster-Shafer Theory of Evidence

In DS theory a piece of information is usually described as a mass function on a frame of discernment.

We first give some basic definitions [Shafer, 1976].

Definition 6.1: *Frame of Discernment*

A set is called a frame of discernment (or simply a frame) if it contains mutually exclusive and exhaustive possible answers to a question. The set is usually denoted as Θ, and it is required that at any time, one and only one element in the set is true.

For instance, if we assume that Emma lives in one of the cities, $city_1, \ldots,$ $city_6$, then

$$\Theta = \{city_1, city_2, city_3, city_4, city_5, city_6\}$$

is a frame of discernment for the question 'In which city does Emma live?'

Definition 6.2: *Mass Function*

A function $m : 2^\Theta \rightarrow [0,1]$ is called a mass function on frame Θ if it satisfies the following two conditions:

1. $m(\emptyset) = 0,$
2. $\Sigma_A m(A) = 1,$

where $\emptyset$ is an empty set and A is a subset of Θ.

A mass function is also called a *a basic probability assignment*, denoted as *bpa*.

For instance, if we know that Emma lives in a area covering the above six cities, but we have no knowledge about in which city she lives, then we can only give a mass function $m(\Theta) = 1$.

Alternatively, if we know that Emma lived in $city_3$ two years ago and she intended to move to other cities and tried to find a job somewhere within these six cities, but we have no definite information about where she lives now, then a mass function could be defined as:

$$m(city_3) = p \qquad \text{and} \qquad m(\Theta) = 1 - p,$$

where p stands for the degree of our belief that she still lives in $city_3$.

Definition 6.3: *Belief Function*

A function $bel : 2^\Theta \to [0, 1]$ *is called a belief function if it satisfies:*

1. $bel(\Theta) = 1$,
2. $bel(\cup_1^n A_l) \geq \Sigma_l bel(A_l) - \Sigma_{l>j} bel(A_l \cap A_j) + \ldots + (-1)^{-n} bel(\cap_l A_l)$.

It is easy to see that $bel(\emptyset) = 0$, for any belief function.

A belief function is also called a *support function.* The difference between $m(A)$ and $bel(A)$ is that $m(A)$ is our belief committed to the subset A excluding any of its subsets while $bel(A)$ is our degree of belief in both A and all its subsets.

In general, if m is a mass function on frame Θ, then

$$bel(A) = \Sigma_{B \subseteq A} m(B). \tag{6.1}$$

is a belief function on Θ.

Recovering a mass function from a belief function is done as follows [Shafer, 1990]:

$$m(A) = \Sigma_{B \subseteq A} (-1)^{|B|} bel(B).$$

For any finite frame, it is always possible to get the corresponding mass function from a belief function and the mass function is unique.

A subset A with $m(A) > 0$ is called a *focal element of this belief function.*

If all focal elements of a belief function are the singletons of Θ, then the corresponding mass function is exactly a probability distribution on Θ. So, mass functions are generalized probability distributions in this sense.

If there is only one focal element of a belief function and the focal element is the whole frame Θ, this belief function is called a *vacuous belief function.* It represents the total ignorance (because of lack of knowledge).

In the following, we call (Θ, bel) a *DS structure.*

Definition 6.4: *Plausibility Function*

A function

$$pls(A) = 1 - bel(\neg A)$$

is called a plausibility function.

A plausibility function $pls(A)$ represents the degree to which the evidence fails to refute A.

From a mass function, we can get its plausibility function as follows [Shafer, 1990]:

$$pls(B) = \Sigma_{A \cap B \neq \emptyset} m(A). \tag{6.2}$$

In a system using evidential reasoning, knowledge or inference results are usually represented by the interval of *bel* and *pls*.

There are several special features of this interval [Wesley, 1983]:

$[bel(A), pls(A)] = [1, 1]$ subset A completely true;

$[bel(A), pls(A)] = [0, 0]$ subset A completely false;

$[bel(A), pls(A)] = [0, 1]$ subset A completely ignorant;

$[bel(A), pls(A)] = [bel, 1]$, $0 < bel < 1$ tends to support A;

$[bel(A), pls(A)] = [0, pls]$, $0 < pls < 1$ tends to refute A;

$[bel(A), pls(A)] = [bel, pls]$, $0 < bel < pls < 1$ may support or refute A.

When more than one mass function is given on the same frame of discernment, then the combined impact of these mass functions is obtained using a mathematical formula called *Dempster's combination rule.*

If m_1 and m_2 are two mass functions on frame Θ, then $m = m_1 \oplus m_2$ is the mass function after combining m_1 and m_2, and

$$m(C) = \frac{\Sigma_{A \cap B = C} m_1(A) m_2(B)}{1 - \Sigma_{A \cap B = \emptyset} m_1(A) m_2(B)},$$

and $\oplus$ means the application of Dempster's combination rule.

The condition of using the rule is stated as "two or more pieces of evidence are based on distinct bodies of evidence" [Shafer, 1976].

6.2 Probability Background of Mass Functions

Even though Shafer has not agreed with the idea that belief function theory is a generalized probability theory and has regarded belief function theory as a new way of representing evidence and knowledge, some people have argued (see, e.g., [Fagin and Halpern, 1989a, Fagin and Halpern, 1989b]) that the theory has strong links with probability theory.

Here, we explore the motivation of Shafer's definition of mass functions under Dempster's original work. We argue that in Dempster's original paper

[Dempster, 1967], he implicitly gave a prototype of mass functions. Shafer's contribution has been to explicitly define the mass function and use it to directly represent evidence.

6.2.1 Dempster's Probability Prototype of Mass Functions

We have the following definitions:

Definition 6.5: *Dempster's Probability Space*

A structure (X, τ, μ) is called a Dempster's probability space where

- *X is a sample space containing all the possible worlds;*
- *τ is a class of subsets of X;*
- *μ is a probability measure which gives $\mu : \tau \to [0,1]$ and $\mu(X) = 1$.*

Definition 6.6: *Multivalued Mapping*

Function $\Gamma : X \to 2^S$ is a multivalued mapping from space X to space S if Γ assigns a subset $\Gamma(x) \subseteq S$ to every $x \in X$.

From a multivalued mapping Γ, a probability measure μ on X can be propagated to space S in such a way that for any subset T of S, the lower and upper bounds of probabilities of T are defined as:

$$P_*(T) = \mu(T_*)/\mu(S^*), \tag{6.3}$$

and

$$P^*(T) = \mu(T^*)/\mu(S^*), \tag{6.4}$$

where:

$$T_* = \{x \in X, \Gamma(x) \neq \emptyset, \Gamma(x) \subseteq T\},$$

and

$$T^* = \{x \in X, \Gamma(x) \cap T \neq \emptyset\}.$$

Equations (6.3) and (6.4) are defined only when $\mu(S^*) \neq 0$. The denominator $\mu(S^*)$ is a renormalizing factor necessitated by the fact that the subset $\{x \mid \Gamma(x) = \emptyset\}$, which does not map into a meaningful subset of S, should be removed from X and the measure of the remaining set S^* renormalized to unity.

A multivalued mapping Γ from space X to space S says that if a possible answer to a question described in the first space X is x, then a possible answer to a question described in the second space S is in $\Gamma(x)$.

Table 6.1: Upper and lower probabilities of all subsets of S

T	$P^*(T)$	$P_*(T)$
$\emptyset$	0	0
$\{s_1\}$	$\frac{1}{k}(p_{100}+p_{110}+p_{101}+p_{111})$	$\frac{1}{k}p_{100}$
$\{s_2\}$	$\frac{1}{k}(p_{010}+p_{110}+p_{011}+p_{111})$	$\frac{1}{k}p_{010}$
$\{s_3\}$	$\frac{1}{k}(p_{001}+p_{101}+p_{011}+p_{111})$	$\frac{1}{k}p_{001}$
$\{s_1, s_2\}$	$\frac{1}{k}(p_{100}+p_{010}+p_{110}+$ $+p_{101}+p_{011}+p_{111})$	$\frac{1}{k}(p_{100}+p_{010}+p_{110})$
$\{s_1, s_3\}$	$\frac{1}{k}(p_{100}+p_{001}+p_{110}+$ $+p_{101}+p_{011}+p_{111})$	$\frac{1}{k}(p_{100}+p_{001}+p_{101})$
$\{s_2, s_3\}$	$\frac{1}{k}(p_{010}+p_{001}+p_{110}+$ $+p_{101}+p_{011}+p_{111})$	$\frac{1}{k}(p_{010}+p_{001}+p_{011})$
S	1	1

For the case when $S = \{s_1, s_2, \ldots, s_n\}$ is finite, the propagation procedure can be done as follows. Suppose $S_{\gamma_1\gamma_2\cdots\gamma_n}$ denotes the subset of S which contains s_l if $\gamma_l = 1$ and excludes s_l if $\gamma_l = 0$ for $l = 1, 2, \ldots, n$.

If for each $S_{\gamma_1\gamma_2\cdots\gamma_n}$, we define $X_{\gamma_1\gamma_2\cdots\gamma_n}$ as

$$X_{\gamma_1\gamma_2\cdots\gamma_n} = \{x \in X, \Gamma(x) = S_{\gamma_1\gamma_2\cdots\gamma_n}\}, \tag{6.5}$$

then all the non-empty subsets of X defined in equation (6.5) form a partition[1] of X and

$$X = \cup_{\gamma_1\gamma_2\cdots\gamma_n} X_{\gamma_1\gamma_2\cdots\gamma_n}. \tag{6.6}$$

The idea of constructing $X_{\gamma_1\gamma_2\cdots\gamma_n}$ is that each $X_{\gamma_1\gamma_2\cdots\gamma_n}$ contains only those elements of X which have the same mapping environment in S.

In order to calculate $P_*(T)$ and $P^*(T)$, Dempster assumed that if each non-empty $X_{\gamma_1\gamma_2\cdots\gamma_n}$ is in τ, then for any $T \subset S$, $P_*(T)$ and $P^*(T)$ are uniquely determined by the 2^m quantities $p_{\gamma_1\gamma_2\cdots\gamma_n}$:

$$p_{\gamma_1\gamma_2\cdots\gamma_n} = \mu(X_{\gamma_1\gamma_2\cdots\gamma_n}). \tag{6.7}$$

We use an example to demonstrate thie idea.

Example 6.1 (From [Dempster, 1967])

Assume that $S = \{s_1, s_2, s_3\}$. Using $p_{\gamma_1\gamma_2\cdots\gamma_n}$, the lower and upper bounds of probabilities of all subsets of S are given in Table 6.2.1.

Here, k is defined as $\mu(S^*) = 1 - p_{000}$.

Given a subset T of S, the corresponding lower and upper subsets in X are known. For instance if $T = S_{110} = \{s_1, s_2\}$, then $T_* = X_{100} \cup X_{010} \cup X_{110}$ and $T^* = X_{100} \cup X_{010} \cup X_{110} \cup X_{101} \cup X_{011} \cup X_{111}$.

[1] A list of subsets $X_1, X_2, \ldots, X_n$ of X is called a partition of X if $X_l \cap X_j = \emptyset$ and $\cup_{l=1}^{l=n} X_l = X$.

Table 6.2: Upper and lower probabilities of all subsets of S using function m

T	$P^*(T) = pls$	$P_*(T) = bel$	$m(T)$
$\emptyset$	0	0	0
$\{s_1\}$	$m_{100} + m_{110} + m_{101} + m_{111}$	m_{100}	m_{100}
$\{s_2\}$	$m_{010} + m_{110} + m_{011} + m_{111}$	m_{010}	m_{010}
$\{s_3\}$	$m_{001} + m_{101} + m_{011} + m_{111}$	m_{001}	m_{001}
$\{s_1, s_2\}$	$m_{100} + m_{010} + m_{110} +$ $+ m_{101} + m_{011} + m_{111}$	$m_{100} + m_{010} + m_{110}$	m_{110}
$\{s_1, s_3\}$	$m_{100} + m_{001} + m_{110} +$ $+ m_{101} + m_{011} + m_{111}$	$m_{100} + m_{001} + m_{101}$	m_{101}
$\{s_2, s_3\}$	$m_{010} + m_{001} + m_{110} +$ $m_{101} + m_{011} + m_{111}$	$m_{010} + m_{001} + m_{011}$	m_{011} x
S	1	1	m_{111}

If we define a function m on S given as $m(S_{\gamma_1\gamma_2\cdots\gamma_n}) = m_{\gamma_1\gamma_2\cdots\gamma_n} = p_{\gamma_1\gamma_2\cdots\gamma_n}/(1 - p_{00\ldots0})$, as in Table 6.2.1, then function m is exactly a mass function when S is a frame. Certainly, some of $m_{\gamma_1\gamma_2\cdots\gamma_n}$ may be 0. P_* and P^* define a belief function and a plausibility function on S respectively.

We assume that this is the model for defining mass functions in Shafer's style. S being a frame of discernment is a special case of S being a space in Dempster's paper.

A vital requirement of calculating probability bounds in Dempster's prototype is that every non-empty subset $X_{\gamma_1\gamma_2\cdots\gamma_n}$ should be in τ. If τ, a collection of subsets of X, does not satisfy this requirement, then the rest of the calculation in Dempster's paper could not be carried out.

6.2.2 Deriving Mass Functions from Probability Spaces

In Chapter 2 we have introduced probability spaces in Definition 2.3. Here we recall it again to refresh our mind.

A structure (X, χ, μ) is called a probability space if (X, χ, μ) is a Dempster's probability space and χ is a σ−algebra of X (a set of subsets of X containing X and closed under complementation and countable union, but not necessarily containing all subsets of X).

A subset χ' of χ is called a *basis* of χ if it contains non-empty and disjoint elements, and if χ consists precisely of countable unions of members of χ'. For any finite χ, there is always a basis of χ, and the basis is unique, i.e.,

$$\Sigma_{X_l \in \chi'} \mu(X_l) = 1.$$

For any subset X_l of X, if X_l is not in χ, it is only possible to get two probability bounds of X_l, usually called an *inner measure*, denoted as μ_*, and an *outer measure*, denoted as μ^*, with:

$$\mu_*(X_l) = sup\{\mu(X_j) \mid X_j \subseteq X_l, X_j \in \chi\}, \tag{6.8}$$

and

$$\mu^*(X_l) = inf\{\mu(X_j) \mid X_j \supseteq X_l, X_j \in \chi\}. \tag{6.9}$$

It was proved in [Fagin and Halpern, 1989a, Fagin and Halpern, 1989b] that μ_* is a belief function on X when X is a frame.

Given a probability space (X, χ, μ), assume there is a multivalued mapping Γ from X to frame S. For a subset T of S, we define

$$bel(T) = \begin{cases} \mu(T_*)/k & \text{when } T_* = \{x \mid \Gamma(x) \neq \emptyset, \Gamma(x) \subseteq T\} \in \chi, \\ \mu_*(T_*)/k & \text{otherwise.} \end{cases} \tag{6.10}$$

bel is also a belief function on S. Here, $k = \mu_*(\{x \in X, \Gamma(x) \cap S \neq \emptyset\})$ which has the same meaning as in Table 6.2.1.

In Dempster's probability space, since those non-empty subsets $X_{\gamma_1\gamma_2\ldots\gamma_n}$ of X form a partition of X, the equation

$$\Sigma_{\gamma_1\gamma_2\ldots\gamma_n}\mu(X_{\gamma_1\gamma_2\ldots\gamma_n}) = 1$$

/noindent holds. This suggests that those non-empty $X_{\gamma_1\gamma_2\ldots\gamma_n}$ possess the properties of a basis of a $\sigma-$algebra. If we use χ' to denote this partition, we can derive a $\sigma-$algebra of X using this basis.

Therefore, Dempster's probability space can always be translated into a normal probability space. So, a probability space is more general than a Dempster's probability space.

Given a probability space (X, χ, μ) and a multivalued mapping Γ from X to S, equation (6.10) is suitable for deriving a belief function on S, while given Dempster's probability space (X, τ, μ) and a multivalued mapping Γ from X to S, equation (6.3) is adequate to define a belief function on S. In either case, a belief function can always be defined, so that belief function theory is closely related to probability theory and comes out of probability theory.

In the following, we use probability spaces to stand for both normal and Dempster's probability spaces.

6.3 Problems with Dempster's Combination Rule

A multivalued mapping from space X_1 to space S generates a mass function on space S given a probability distribution on X or a collcction of subsets of X_1 (either τ or χ) as we have seen in the previous section.

If there is another multivalued mapping from space X_2 to S which also defines a mass function on S, then there are two mass functions available on S. In DS theory, when there are two or more mass functions on the same frame of discernment, it is desirable to combine them using Dempster's combination rule, if it is believed that they are derived from 'independent sources'.

The debate about the precise meaning of 'independent sources' has appeared in many research papers, including some by Shafer himself, since the interpretation of this phase determines whether an agent can apply Dempster's rule correctly.

However, Dempster's combination rule only involves mass functions which cannot tell anything about a relationships with their sources. Therefore, it is very difficult to portray the condition of using the rule if an agent's attention is focused on the rule itself.

As Dempster's combination rule is motivated by the idea in Dempster's original paper, we believe that the re-examination of the idea of combination suggested by Dempster would be very beneficial for the clarification of the condition of using the rule.

6.3.1 Dempster's Combination Framework

In [Dempster, 1967], Dempster first discussed the procedure of generating a lower and upper bounds of probabilities on space S from a probability space (X, χ, μ) through a multivalued mapping Γ, as discussed in Section 6.2. Then, Dempster considered the situation when n probability spaces are available and each of which has a multivalued mapping with the same space S, given that these n probability spaces are independent.

In summary, Dempster's idea on the combination of independent sources of information can be stated as follows.

Suppose there are n pieces of evidence which are given in the form of n probability spaces, (X_l, χ_l, μ_l), each of which has a mapping relationship with the same space S through a multivalued mapping Γ_l.

These n sources are said to be independent and explained by Dempster as

> "...opinions of different people based on overlapping experiences could not be regarded as independent sources. Different measurements by different observations on different equipment would often be regarded as independent ...the sources are statistically independent [Dempster, 1967]".

Under his assumption of independence, Dempster [Dempster, 1967] suggested that these sources could be combined using equation (6.11) to derive a combined source (X, χ, μ) and a unified Γ as:

$$X = X_1 \otimes X_2 \otimes \ldots \otimes X_n,$$

$$\chi = \chi_1 \otimes \chi_2 \otimes \ldots \otimes \chi_n, \tag{6.11}$$

$$\mu = \mu_1 \otimes \mu_2 \otimes \ldots \otimes \mu_n,$$

and

$$\Gamma(x) = \Gamma_1(x) \cap \Gamma_2(x) \cap \ldots \cap \Gamma_n(x).$$

The fourth formula can also be stated and explained as

$$\Gamma(x) = \Gamma'_1(x) \cap \Gamma'_2(x) \cap \ldots \cap \Gamma'_n(x)$$

where: $\Gamma'_l(x) = \Gamma_l(x_l)$, when $x \in X_1 \otimes \ldots \otimes X_{l-1} \otimes \{x_l\} \otimes \ldots \otimes X_n$.

Here, $\otimes$ is the set product operator as usual, and $\mu = \mu_1 \otimes \mu_2$ is defined as $\mu(\{(x_{1l}, x_{2j})\}) = \mu_1(\{x_{1l}\}) \times \mu_2(\{x_{2j}\})$, for every $(x_{1l}, x_{2j}) \in X$, where $x_{1l} \in X_1$ and $x_{2j} \in X_2$.

The combined source (set product measure space) summarises the message carried by all separate sources. The combined multivalued mapping Γ from the combined source to space S suggests that $x = (x_1, x_2, \ldots, x_n) \in X$ is consistent with $s_l \in S$ iff s_l belongs to all $\Gamma_l(x_l)$ simultaneously.

Through the combined source (X, χ, μ) and the unified mapping Γ, for any subset T of S, a pair of lower and upper bounds of probabilities can be calculated using equations (6.3) and (6.4).

Apart from using the set of formulae listed in equation (6.11), Dempster also recommended an alternative calculation method for calculating the final bounds of probabilities on S using the function

$$q(T) = q_1(T) \times q_2(T) \times, \ldots, \times q_n(T), \tag{6.12}$$

where

$$q_l(T) = \mu_l(\tilde{T}_l),$$

and

$$\tilde{T}_l = \{x_l \in X_l, \Gamma_l x_l \supseteq T\}.$$

When S is finite, each q_l can be calculated through $p^l_{\gamma_1\gamma_2\ldots\gamma_n}$. For instance, $q^l_{100} = p^l_{100} + p^l_{110} + p^l_{101} + p^l_{111}$ when S has three elements.

If we examine equations (6.11) and (6.12) carefully, then it is easy to see that these two different calculation methods suggest two different procedures of combination. The former calculates the combined source first while the latter propagates the effect of each source to S first.

Dempster implicitly assumed that the results obtained in these two procedures were the same under the condition that the n sources were *statistically independent*.

If we refer to the levels containing spaces (X_l, χ_l, μ_l) as the *original information level*, and space S as the *target information level*, then Dempster's condition of independence is assumed at the original information level. This requirement is called *DS-independent* in [Voorbraak, 1991].

The two approaches in Dempster's combination framework can be summarised as follows:

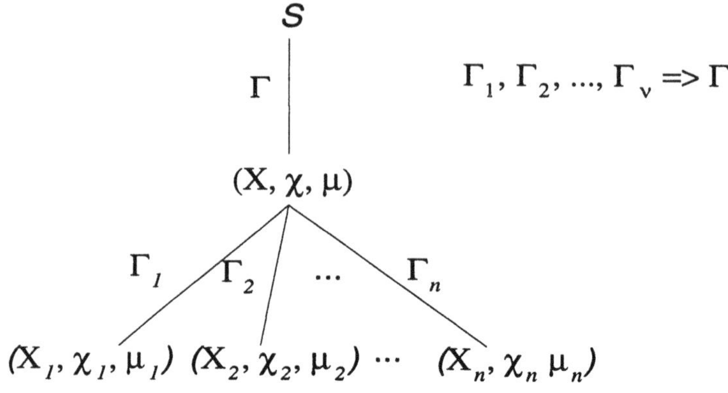

Figure 6.1: Find a common probability space first and then propagate the combined probabilities distribution to the target space

Approach 1: Combining information at the original information level by producing a joint space and a single probability distribution on the space.

This procedure should consider the different mappings from the joint space to the target information space, unify the original mappings into one mapping and propagate the joint probability distribution to the target information level.

The intuitive meaning of this method is shown in Figure 6.1.

Approach 2: Propagating different pieces of evidence at the original information level to the target information level separately and then combining them at the target information level.

Figure 6.2 illustrates this procedure explicitly.

6.3.2 The Condition for Using Dempster's Combination Rule

Proposing Dempster's combination rule in his book, Shafer [Shafer, 1976] followed the spirit of *Approach 2* suggested by Dempster, where function q is replaced by function m in Figure 6.2. On S, Shafer proposed a mathematical formula called Dempster's combination rule to combine these mass functions.

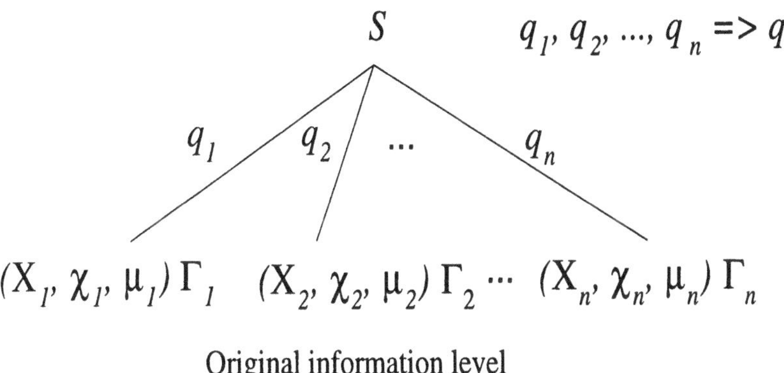

Figure 6.2: Propagate evidence individually to the target space first and then carry out combination at the target information level

If we follow the idea that the rule proposed by Shafer was derived from Dempster's combination framework, (i.e., Approach 2), then Dempster's combination rule should obey the above condition defined by Dempster.

Later, in some papers, Shafer indeed addressed the importance of independence among original sources. For example in [Shafer, 1986, Shafer, 1987], Shafer stated that the condition of using Dempster's combination rule is that 'two or more belief functions on the same frame but based on independent arguments or items of evidence' and in [Shafer, 1982] he used randomly encoded messages to describe the condition of using Dempster's combination rule.

These explanations are much closer to the definition given by Dempster in his combination framework. However, as Dempster's combination rule does not require or reflect any information about the sources which support the corresponding belief functions and it only needs belief functions (or mass functions) in order to carry out the combination, it is, therefore, difficult to describe independent conditions precisely using only belief functions.

In contrast to two views on belief functions in [Halpern and Fagin, 1992], we argue that the main cause of giving counterintuitive results[2] in using

[2]These do not include the situations such as Example 6.12 in [Voorbraak, 1991] and the murderer case in [Smets, 1988], Section 5.2. Example 6.12 in [Voorbraak, 1991] demonstrates that Dempster's combination rule agrees that a mass value assigned to a subset should not be evenly split among its elements but be kept for all the elements in this subset.

The murderer case in Section 5.2 in [Smets, 1988] suggests that the normalisation function of Dempster's combination rule may not produce reasonable results when two pieces

Dempster's combination rule is *overlooking* or (*ignorance*) of the condition of combination given in Dempster's original paper.

In Shafer's simplified combination mechanism (i.e., Dempster's combination rule) the original sources are hidden. The invisibility of the original sources in the simplified combination mechanism makes it difficult to state the condition of the mechanism.

In other words, Dempster's combination rule is too simple (compared to Dempster's combination framework) to show (or carry) enough information for a precise mathematical description of the dependent (or independent) relations between multiple pieces of evidence.

We have now the following proposition:

Proposition 6.1:

Two belief functions on a frame can be combined using Dempster's combination rule if the two sources, (X_1, χ_1, μ_1) and (X_2, χ_2, μ_2), from which the two belief functions are derived, are statistically independent (or called DS-independent).

Shafer [Shafer, 1982], Shafer and Tversky [Shafer and Tversky, 1985] and Voorbraak [Voorbraak, 1991] also mentioned the idea of describing and judging dependent relations among the original probability spaces but not explicitly defined it at the original information level.

6.3.3 Examples

In this section, we use three examples to illustrate a delicate difference between the two approaches in Dempster's combination framework. The first two examples come from [Shafer, 1986] where the former shows that when two pieces of evidence are statistically independent, both approaches in Dempster's combination framework are applicable, and the latter shows that when they are not statistically independent, only the first approach works. The third example ([Voorbraak, 1991]) not only demonstrates that a careless application of Dempster's combination rule will yield a wrong result but also reveals the possibility of using *Approach 1* to combine two dependent pieces of evidence when their common probability space is known (as also investigated in [Shafer, 1986], [Shafer, 1987], [Lingras and Wong, 1990]), although Dempster's original intention aimed at dealing with independent probability spaces.

Example 6.2

Suppose that Shafer wants to know whether the street outside is slippery, instead of observing this himself, he asks another person, named Fred. Fred

of evidence are almost contradictory with each other. These examples are not within the scope of this book.

tells him that 'it is slippery'. However Shafer knows that Fred is sometimes careless in answering questions. Based on his knowledge about Fred, Shafer estimates that 80% of the time Fred reports what he knows and he is careless 20% of the time. So, Shafer believes that there is only an 80% chance that the street is slippery.

In fact, Shafer forms two frames X_1 and S in his mind in order to get his conclusion in this problem, where X_1 is related to Fred's truthfulness and S is related to the possible answers of slippery outside:

- $X_1 = \{truthful, careless\}$,
- $S = \{yes, no\}$,

and, here, *yes* and *no* stand for 'it is slippery' and 'it is not slippery', respectively.

A probability measure p_1 on X_1 is defined as:

$$\mu_1(\{truthful\}) = 0.8,$$

and

$$\mu_1(\{careless\}) = 0.2.$$

Shafer derives his conclusion when this probability measure is propagated from frame X_1 to frame S through a multivalued mapping function between X_1 and S as:

$$\Gamma_1(truthful) = \{yes\},$$

and

$$\Gamma_1(careless) = \{yes, no\}.$$

Shafer obtains a belief function on S, based on equation (6.10), as:

$$bel_1(\{yes\}) = 0.8, \qquad bel_1(\{no\}) = 0.0. \tag{6.13}$$

Furthermore, suppose that Shafer has some other evidence about whether the street is slippery: his trusty indoor-outdoor thermometer says that the temperature is 31 degrees Fahrenheit, and he knows that because of the traffic ice could not form on the streets at this temperature.

However he knows that the thermometer could be wrong even though it has been very accurate in the past. Suppose that there is a 99% chance that the equipment is working properly, so he could form another frame X_2 with its probability distribution as:

$$X_2 = \{working, not\_working\},$$

and

$$\mu_2(\{working\}) = 0.99, \qquad \mu_2(\{not\_working\}) = 0.01,$$

and a mapping function Γ_2 as

$$\Gamma_2(working) = \{no\},$$

and

$$\Gamma_2(not\_working) = \{yes, no\}.$$

Therefore, another belief function on S is calculated, using equation (6.10), as:

$$bel_2(\{yes\}) = 0.0, \qquad bel_2(\{no\}) = 0.99. \tag{6.14}$$

Now, there are two pieces of evidence available regarding the same question 'slippery or not?' and Shafer wants to know what the joint impact of the evidence on S is.

In the following, we try to solve this problem for Shafer in Dempster's combination framework.

Using Approach 1 in Dempster's combination framework:

First of all, there are two probability spaces $(X_1, 2^{X_1}, \mu_1)$ and $(X_2, 2^{X_2}, \mu_2)$ carrying the original information, which are believed to be independent as Fred's answer is independent of the output of the equipment.

According to Dempster's explanation about independence of sources, equation (6.11) is applied to combine these two sources to obtain a combined source and a joint multivalued mapping.

As a result, the combined source is $(X, 2^X, \mu)$ where

$$X = X_1 \otimes X_2$$

and

$$\mu(\{(x_1, x_2)\}) = \mu_1(\{x_1\}) \times \mu_2(\{x_2\}),$$

such that

$$\begin{aligned}\mu(\{(truthful, working)\}) &= \mu_1(\{truthful\}) \times \mu_2(\{working\}) \\ &= 0.8 \times 0.99 = 0.792,\end{aligned}$$

as shown in the second column in Table 6.3.3.

In the meanwhile, the joint multivalued mapping Γ from X to S is defined as

$$\begin{aligned}\Gamma((truthful, working)) &= \emptyset, \\ \Gamma((truthful, not)) &= \{yes\}, \\ \Gamma((careless, working)) &= \{no\}, \\ \Gamma((careless, not)) &= \{yes, no\},\end{aligned}$$

Table 6.3: The combined source is simply the set product of the original sources

$X = X_1 \otimes X_2$ combined source	$\mu = \mu_1 \otimes \mu_2$ joint probability	$\Gamma(x) \subseteq S$ joint mapping
(truthful, working)	0.792	$\emptyset$
(truthful, not)	0.008	{yes}
(careless, working)	0.198	{no}
(careless, not)	0.002	{yes, no}

after taking into account both what Fred and the thermometer have said. Here, 'not' means 'not_working'. Element (truthful, working) is the only element which matches the empty set in S, so that $\mu(S^*) = 1 - 0.792 = 0.208$.

As a result, using equation (6.3), we get

$$\begin{aligned} & P_*(\{yes\}) = \\ = \ & \mu(\{yes\}_*)/\mu(S^*) = \\ = \ & \mu(\{(truthful, not)\})/\mu(S^*) = \\ = \ & 0.008/0.208 = 0.04. \end{aligned}$$

and this is the degree of our belief that the road is indeed slippery.

Alternatively, it is also possible to use equation (6.10) to calculate the degree of our belief

$$bel(\{yes\}) = 0.04,$$

which is the same as P_*.

Using Approach 2 in Dempster's combination framework:

In this approach, function q is to be used to calculate the joint impact on S without constructing the joint probability space, according to equation (6.12).

Given two original probability spaces $(X_1, 2^{X_1}, \mu_1)$ and $(X_2, 2^{X_2}, \mu_2)$ as defined in the first part of this example, it is possible to calculate q on S as:

$$\begin{aligned} & q(\emptyset) = q_1(\emptyset) \times q_2(\emptyset) = 1 \times 1 = 1, \\ & q(\{yes\}) = q_1(\{yes\}) \times q_2(\{yes\}) = 1 \times 0.01 = 0.01, \\ & q(\{no\}) = q_1(\{no\}) \times q_2(\{no\}) = 0.2 \times 1 = 0.2, \\ & q(S) = q_1(S) \times q_2(S) = 0.2 \times 0.01 = 0.002. \end{aligned}$$

Dempster also suggested that when S is finite, then q can be expressed in the form of $p_{\gamma_1, \ldots \gamma_n}$.

In this example, S contains only two elements, it is therefore possible to re-write function q as follows:

Table 6.4: The direct application of Dempster's combination rule

m		$\{yes\}$	0.8	$\{yes, no\}$	0.2
$\{no\}$	0.99	$\emptyset$	0.792	$\{no\}$	0.198
$\{yes, no\}$	0.01	$\{yes\}$	0.008	$\{yes, no\}$	0.002

$$\begin{aligned}
&q(\emptyset) = q_{00} = p_{00} + p_{10} + p_{01} + p_{11} = 1,\\
&q(\{yes\}) = q_{10} = p_{10} + p_{11} = 0.01,\\
&q(\{no\}) = q_{01} = p_{01} + p_{11} = 0.2,\\
&q(S) = q_{11} = p_{11} = 0.002.
\end{aligned}$$

This set of equations determines $p_{\gamma_1\gamma_2}$ with the following values:

$$p_{00} = 0.792, \qquad p_{10} = 0.008,$$

and

$$p_{01} = 0.198, \qquad p_{11} = 0.002.$$

Therefore, using the method shown in Table 6.2.1 or Table 6.2.1, the degree of our belief in the statement 'the outside is slippery' is

$$P_*(\{yes\}) = p_{10}/(1 - P_{00}) = 0.008/0.208 = 0.04 = bel(\{yes\}).$$

This result is consistent with what has been obtained in *Approach 1.*

Alternatively, as we believe that Dempster's combination rule proposed by Shafer is inspired by and a variation of equation (6.12), then it is also possible to apply the rule to two belief functions in (6.13) and (6.14) on S directly. The combined result is shown in Table 6.3.3.

The first row and the first column stand for the two mass functions derived from the two belief functions defined in equations (6.13) and (6.14), respectively.

The combined mass function on yes is $m(\{yes\}) = 0.008/0.208 = 0.04$, so that $bel(\{yes\}) = 0.04$ which is identical to the result in both approaches.

◊

Let us not forget Shafer's initial independence assumption between Fred's opinion and the output of the thermometer when we attempted to solve this problem.

Next we will see when this assumption no longer holds, what results these approaches will offer.

Table 6.5: The combined space and its new probability distribution from the two original spaces, when the original two spaces are not independent

$X = X_1 \otimes X_2$ combined source	μ' joint probability	$\Gamma(x) \subseteq S$ joint mapping
$(truthful, working)$	0.799	$\emptyset$
$(truthful, not)$	0.001	$\{yes\}$
$(careless, working)$	0.191	$\{no\}$
$(careless, not)$	0.009	$\{yes, no\}$

Example 6.3

Continuing Example 6.2, assume that Shafer believes that the Fred's answer relates to the thermometer as Fred looks at the thermometer regularly to see whether it is working. If it is not working properly, then Fred would be careless in answering questions. Assume that Fred has a 90% chance of being careless if the thermometer is not working, then Fred's answer is somehow affected when the thermometer is not working.

Using Approach 1 in Dempster's combination framework:

Because the two original probability spaces are not statistically independent, it is impossible to get the joint probability distribution μ on the joint space X by simply applying $\otimes$ to μ_1 and μ_2. Under these circumstances, Shafer somehow worked out an alternative joint probability distribution μ' on the same joint space X ($X = X_1 \otimes X_2$). The probability that μ' assigns on each element in X is shown in the second column in Table 6.3.3. Based on this newly created joint probability space (X, χ, μ') and the original multivalued mapping Γ between X and S (shown in the third column in Table 6.5), equation (6.3) [or (6.10)] once again is used to calculate the degree of our belief:

$$\begin{aligned} P_*(\{yes\}) &= \mu'(\{yes\}_*)/\mu'(S^*) = \mu'(\{(truthful, not)\})/\mu'(S^*) \\ &= 0.001/0.201 = 0.005 = bel(\{yes\}), \end{aligned}$$

where $\mu'(S_*) = 1 - 0.799$.

This result is obviously different from the result given in Example 6.2 because of the relationship between the two pieces of evidence.

Using Approach 2 in Dempster's combination rule:

If we took a chance and believed that Approach 2 in Dempster's combination framework could be applied, then we would end up with exactly the same calculation procedure, and the same result of course, as in Example 6.2,

Table 6.6: Comparison of Dempster's original combination framework and Dempster's combination rule in different situations

	Dempster's combination framework		Dempster's combination rule
	Approach 1	Approach 2	bel_1 bel_2
Ex. 6.2	Applicable and correct	Applicable and correct	Applicable and correct
Ex. 6.3	Applicable and correct	Inapplicable	Inapplicable

because the two original sources are still the same. When function q is being calculated, there are no requirements on the joint probability of the originial sources, so the change Shafer made in μ' has no way to be reflected.

Therefore, the second approach should not be used in the non-independent situation at all. As a straightforward consequence, Dempster's combination rule should not be allowed to apply in this case as well[3], as the rule is practically a different way of calculating q.

A summary of this analysis is shown in Table 6.3.3.

We can see from Table 6.3.3 that whenever *Approach 2* is applicable, Dempster's rule is guaranteed to be applicable and whenever *Approach 2* is **not** applicable, Dempster's combination rule **is not** applicable either.

If an agent uses Dempster's combination rule under the condition that the rule comes from Dempster's combination framework, then the agent can normally use the rule correctly. However, if an agent does not have Dempster's combination framework in mind when he intends to apply Dempster's combination rule, then it is sometimes very difficult to judge whether the rule is applicable, given two belief functions.

We use the next example to further emphasize the importance of bearing in mind the fact that Dempster's combination rule comes from *Approach 2* in Dempster's combination framework whenever Dempster's combination is being used.

Examples 5.2 and 5.3 are simplified versions of Examples 6.2 and 6.3, respectively. If we compare these two pairs of examples (Example 5.2 with 6.2, and Example 5.3 with 6.3), then it is not difficult to see that the combination rule in generalized incidence calculus can deal with both cases, but Dempster's rule can only deal with the first.

[3]If Dempster's combination rule were applied, the result would still be $bel(\{yes\}) = 0.4$, which would have not shown the inter-relationship between the two mass functions.

Example 6.4 (from [Voorbraak, 1991])

There are 100 labelled balls in an urn. Each ball must have either label a or b, not both, in addition to some extra labels, x or y or xy.

Agents A and B give separate observations of drawing a ball from the urn as follows:

Agent A: The drawn ball has label x. The space of labels describing those balls is $X_1 = \{axy, ax, bxy, bx\}$ with probability distribution:

$$\mu_1(\{axy\}) = \mu_1(\{ax\}) = 4/28,$$

and

$$\mu_1(\{bxy\}) = \mu_1(\{bx\}) = 10/28.$$

The probability $\mu_1(\{axy\}) = 4/28$ means that a drawn ball has label axy with probability 4/28 when it is known that the label definitely contains x.

Agent B: The drawn ball has label y. The space of labels describing those balls is $X_2 = \{axy, ay, bxy, by\}$ with probability distribution:

$$\mu_2(\{axy\}) = 4/50, \qquad \mu_2(\{ay\}) = 16/50,$$

and

$$\mu_2(\{bxy\}) = 10/50 \qquad \mu_2(\{by\}) = 20/50.$$

The probability $\mu_2(\{axy\}) = 4/50$ means that a drawn ball has label axy with probability 4/50 when it is known that the label definitely contains y.

Based on these two pieces of evidence, we are interested in knowing the degree of our belief that the drawn ball also has label b.

First of all, we try to work out the solution using Dempster's combination rule.

Using Dempster's combination rule:

Let $\{a, b\}$ be a frame of discernment, where a stands for 'the drawn ball has label a' and b stands for 'the drawn ball has label b'.

Two mass functions are defined on S based on the information carried by two agents A and B as:

$$m_X(a) = 2/7, \quad m_X(b) = 5/7,$$

and

$$m_Y(a) = 2/5, \quad m_Y(b) = 3/5,$$

Table 6.7: All the possible labels, the number of balls with each possible label, the prior probability of drawing a ball having a particular label

Set of labels X	Number of balls having that label	μ
axy	4	0.04
ax	4	0.04
ay	16	0.16
a	16	0.16
bxy	10	0.1
bx	10	0.1
by	20	0.2
b	20	0.2

where $m_X(a)$ is the mass value on a given by agent A's observation which represents the possibility of a ball having label a when the ball is observed having label x and $m_Y(a)$ is the mass value on a given by agent B's observation which represents the possibility of a ball having label a when the ball is observed having label y.

The result of applying Dempster's combination rule to m_X and m_Y is $m(b) = m_X \oplus m_Y(b) = 15/19$. So, $bel(b) = 15/19$.

Before judging whether this result is right or wrong, let us see what results the two approaches in Dempster's combination framework and Bayesian probability can offer.

Using Bayesian probability:

In probability theory, the full information of probability distribution on every possible label must be known. This information is provided in Table 6.3.3.

The probability that a ball has both labels x and y is

$$p(x)p(y) = 0.28 \times 0.5 = 0.14 = p(x \wedge y).$$

Therefore, the conditional probability that a drawing ball also has label b is $p(b \mid x \wedge y) = 5/7$ when both labels x and y are observed.

Using Approach 1 in Dempster's combination framework:

In fact, the information provided by agents A and B are carried by two probability spaces (X_1, χ_1, μ_1) and (X_2, χ_2, μ_2) with $\chi_1 = 2^{X_1}$ and $\chi_2 = 2^{X_2}$, respectively.

Let us construct another space containing labels a and b only, $S = \{a, b\}$. Then, there can be two multivalued mappings between the two spaces (provided by A and B) and S as,

$$\Gamma_1(axy) = \Gamma_1(ax) = \{a\},$$

Table 6.8: Possible labels after A and B's opinions are considered, the number of balls having each possible label, the posterior probability distribution, and a multivalued mapping between X and S

Set of labels X	Number of balls having that label	μ'	$\Gamma(\bullet) \subseteq S$
axy	4	4/14	a
bxy	10	10/14	b

$$\Gamma_1(bxy) = \Gamma_1(bx) = \{b\},$$

$$\Gamma_2(axy) = \Gamma_2(ay) = \{a\},$$

and

$$\Gamma_2(bxy) = \Gamma_2(by) = \{b\}.$$

The idea of *Approach 1* is to get a combined space first before the combined probability is propagated to space S. In this case, these two probability spaces are not independent as they are from the same original probability space.

Agents A and B have only provided partial information. So, the combined probability distribution on the combined space $X = \{axy, bxy\}$ is not $\mu_1 \otimes \mu_2$ but the posterior probability distribution $\mu'(\bullet) = \mu(\bullet/x \wedge y)$ as given in column 3 in Table 6.3.3, after agents A and B's opinions are considered. The combined multivalued mapping function is detailed in the fourth column in Table 6.3.3.

Therefore, according to equation (6.3), the degree of our belief that the drawing ball also has label b is $P_*(\{b\}) = \mu'(\{b\}_*)/\mu'(S^*) = (10/14)/1 = 5/7$, when agents A and B confirmed that the label of a drawing ball contains both x and y.

Using Approach 2 in Dempster's combination framework:

This is a typical situation where the two original probability spaces are not independent, so *Approach 2* cannot be used.

Discussion:

Obviously the result obtained in DS theory is different from that obtained in probability theory and that in Dempster's combination framework, and the result given in DS theory is wrong. The very reason of this wrong result is that since the two pieces of evidence are not statistically independent, then *Approach 2* cannot be applied. Therefore, Dempster's combination rule should not be applied either.

The importance of considering relations among the original information sources has once again been discussed above. The result tells us that it is more natural to consider the combination at both the original information level and the target information level than only at the target information level.

6.4 Computational Complexity Problem in DS Theory

Soon after DS theory was used in practice, it was pointed out that the use of Dempster's combination rule is of a high computational complexity. Some algorithms have been developed to reduce that computational complexity.

6.4.1 Linear Algorithms of Dempster's Combination Rule

We start with the following definition:

Definition 6.7: *Simple Support Function*

A belief function bel on frame Θ is called a simple support function if there is one and only one focal element (other than frame Θ) of this belief function.

For example, *bel* calculated from $m(A) = 0.7$, $m(\Theta) = 0.3$ is a simle support function, where $A \subset \Theta$.

It is clear that a simple support function has only one focal element which should not be the whole frame.

Barnett's algorithm

Barnett [Barnett, 1981] first considered the reduction of the computational complexity of Dempster's rule of combination. He designed an algorithm in which every piece of evidence can be represented as a special case of a simple support function, such that a focal element is either a singleton or the complement of a singleton.

Mathematically, a mass function m which can be considered in Barnett's algorithm is in the form:

$$m(A) \neq 0,$$

iff $A = \{a\}$ or $A = \Theta \setminus \{a\}$, and

$$m(\Theta) = 1 - m(A).$$

Barnett's algorithm could combine l mass functions on a frame with n elements in a linear time with n. For more details, see [Barnett, 1981].

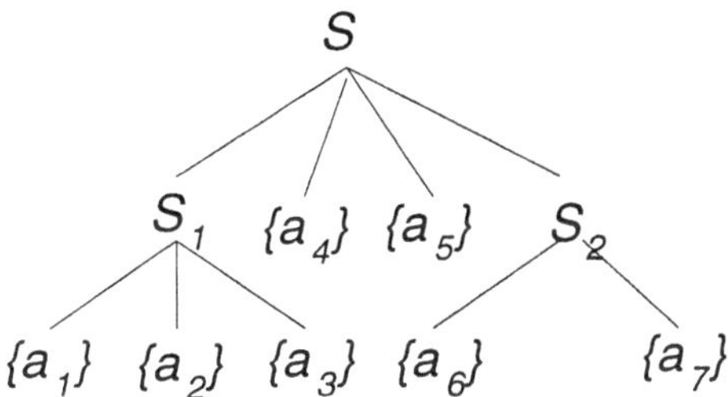

Figure 6.3: A hierarchical evidence (hypothesis) space

Gordon and Shortliffe's algorithm

Barnett's algorithm is not applicable when a mass function is not in the form required. Gordon and Shortliffe [Gordon and Shortliffe, 1985] explored the problem in a slightly more general situation when a simple support function can either be in favor of, or against a subset (not just a singleton) of a frame. However, these subsets must correspond to hypotheses, in a *hierarchical hypothesis space*, which can be arranged as nodes of a tree, like Figure 6.3.

They suggested an approximation of Dempster's combination rule for these nodes, because they argued that in a medical diagnostic system, only these hypotheses matter. Their algorithm carries out in a linear time with the number of leaf nodes in a tree.

Shafer and Logan's algorithm

Shafer and Logan [Shafer and Logan, 1987] extended the results presented in the work [Gordon and Shortliffe, 1985] using an exact implementation of Dempster's rule for the *hierarchical evidence space.* A space of hierarchical evidence is also represented as a tree as shown in Figure 6.3, with nodes standing for hypotheses and evidence is assigned to nodes (or their complements) only. Shafer and Logan chose the term a *hierarchical evidence space* instead of a *hierarchical hypothesis space*, because they wanted to make clear that in their algorithm one could also calculate degrees of belief in other hypotheses (non-nodes) as well.

In order to take advantage of Barnett's algorithm, Shafer and Logan apply the idea of partition on every level of a tree, to form suitable frames. For example, $\{a_1, a_2, a_3, \neg S_1\}$, and $\{S_1, \neg S_1\}$ (where $\neg S$ is the complement of subset S) are two possible frames for calculating evidence assigned to $\{a_1\}$, $\{a_2\}$, $\{a_3\}$ or S_1, or the complements of these elements using Barnett's method. Their algorithm propagates evidence and calculates beliefs with complexity being linear in the total number of nodes in a tree.

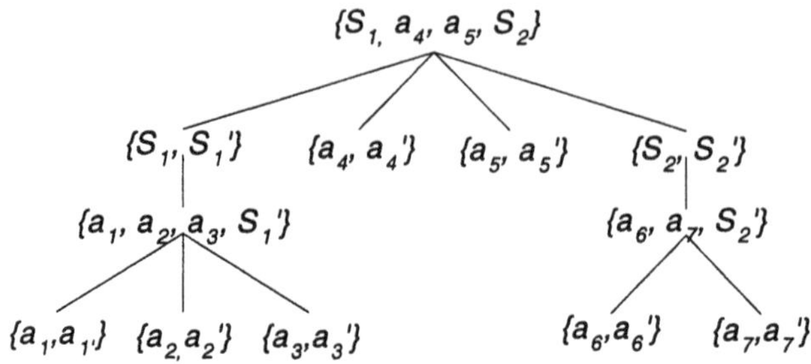

(a) The tree of families and dichotomies

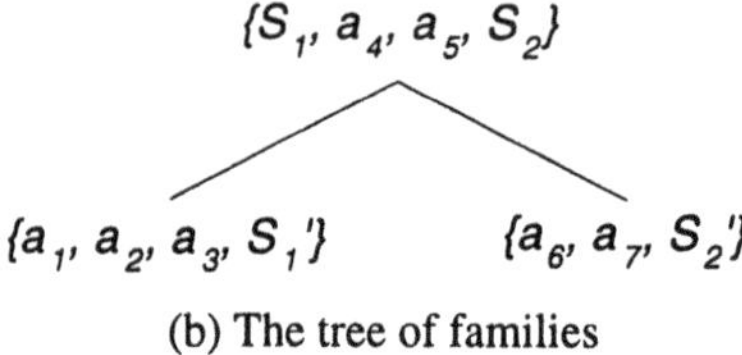

(b) The tree of families

a_1' *stands for the complement of subset* $\{a_1\}$.

Figure 6.4: Two types of qualitative Markov trees from the hierarchical evidence space in Figure 6.3

6.4.2 Algorithms Based on Markov Trees

Shafer and Logan's algorithm [Shafer and Logan, 1987] is inapplicable if the focal elements of a belief function are not nodes or complements of nodes of a hierarchical evidence space. For instance, a mass function with $m(\{a_1, a_2\}) = v_1$ and $m(a_3) = v_2$, where $v_1 + v_2 = 1$, cannot be dealt with in their algorithm.

In [Shafer *et al.*, 1987], the authors use a *qualitative Markov tree*, instead of a diagnostic tree, to bear the evidence. Each node in a qualitative Markov tree is a frame in its own right. Therefore, every belief function, not necessarily a simple support function, is associated with a particular frame and that frame is represented as a node.

For example, from the hierarchical tree in shown in Figure 6.3, two qualitative Markov trees can be built as shown in Figure 6.4.

As far as the computational complexity is concerned, this algorithm reduces the exponential complexity in the size of a frame Θ to the size of the largest node in a tree, a node with the largest number of elements. The authors also concluded that the methods in [Shafer and Logan, 1987], [Shafer, 1985], and [Pearl, 1986] are special cases of the qualitative Markov tree method.

In order to gain more efficiency when a problem domain is very large, research efforts have also been focused on *local computations* of marginals of hypotheses on *Markov trees* (or hypertrees) (we refer the reader to, e.g., the works of [Shenoy and Shafer, 1986], [Lauritzen and Spiegelhalter, 1988], [Shafer and Shenoy, 1988], [Shenoy and Shafer, 1990], etc.) and on *binary join trees* ([Shenoy, 1997], etc.). Each hypothesis is a subset of a frame (or even a frame itself) represented by a node.

The term Markov tree, borrowed from probability theory and first used in [Shafer *et al.*, 1987] as a qualitative Markov tree, means a tree of variables in which a separation implies probabilistic conditional independence given the separating variables [Shenoy and Shafer, 1990]. Markov trees are also called *join trees* in database theory [Maier, 1983]. The other names for Markov trees are, for instance, *junction trees* [Jensen *et al.*, 1990] and *clique trees* [Lauritzen and Spiegelhalter, 1988].

Local computation, which was initiated for propagating probabilities in Bayesian causal trees by Pearl [Pearl, 1986], refers to computation which involves only a small number of nodes in a large tree (or network). The basic idea of local computation is message passing among neighbouring nodes in a Markov tree to compute marginals of the joint belief distribution without actually calculating the joint belief distribution. The sizes of nodes in a Markov tree determine how efficient the local computation can be. If a node has a large number of neighbours, even local computation can be very inefficient. To solve this problem, in [Shenoy, 1997] a concept called a *binary join tree* was proposed in which every non-leaf node has at most two children. Any Markov tree can be first transformed into a binary join tree, and then one can carry out local computation. It is believed that in this way the computation in a Markov tree can be efficient [Shenoy, 1997].

In [Xu, 1995], a situation is considered when it is necessary to compute the marginal of a hypothesis which contains elements from two or more different frames (nodes) rather than elements from a single frame (node). In mathematical terms, there was presented there a technique for computing the marginal for a subset which is not contained in one vertex, but is contained in the union of several vertices.

In [Wilson, 1991], a Monte Carlo algorithm is designed for the calculation of belief in DS theory which performs calculation in time linear in the size of a frame and the total number of mass functions, given that the weight of conflict is bounded. This algorithm can also be used to improve the complexity of the Shenoy-Shafer algorithms on Markov trees.

Propagating belief functions in networks of variables has also been studied by Kong [Kong, 1987] and Mellouli [Mellouli, 1987]. A different approach for reducing the complexity of Dempster's combination rule was studied using the Möbius transform of a graph in [Kennes, 1992] and [Kennes and Smets, 1990].

Another stream of effort has been made on approximating Dempster's combination rule. Voorbraak in [Voorbraak, 1988] defines a Bayesian approximation of a belief function and shows that combining Bayesian approximations of belief functions is computationally less complex than combining belief functions. A consonant approximation method for belief functions has been studied in [Dubois and Prade 1990a]. Recently, Tessem [Tessem, 1993] provided another approximation using the technique of reducing the number of focal elements, and also an evaluation of these three approximation approaches in terms of the error they make and the explanation facilities they provide.

6.4.3 Parallel Techniques for Managing Complexity

Parallel techniques have been proved useful in obtaining a higher efficiency in many aspects. In [Shafer *et al.*, 1987], an ideal situation using parallel techniques is described in which each node in a Markov tree is assigned to a distinct processor. Therefore, the message passing procedure among nodes can be taken as receiving messages from and sending messages to neighbour processors. Each node, now a processor, can perform local computation as soon as it receives necessary inputs.

Wong and Hwang in [Wong and Hwang, 1993] first implemented Dempster's combination rule in a parallel environment where all subsets of frame Θ are evenly divided into a number of groups and each group is assigned to a distinctive processor. When two pieces of evidence (m_1, m_2) are ready to be combined, the control processor will pass these two mass functions to every other processor (slave processor). All the processors react simultaneously. Each processor will then calculate the combined mass value for every subset in that group and pass the result to the control processor.

If there is another piece of evidence m_3, the result of $m_1 \oplus m_2$ is treated as one mass function, and it will be combined with m_3 in the above procedure. This process continues until all the evidence has been dealt with. Then, the control processor will pass the final combined mass function to every slave processor to calculate the degree of belief for every subset.

Apart from this two-level information flow structure, Wong and Hwang also considered the parallel process of evidence on a tree with more than two levels. This tree is very similar to the hierarchical hypothesis space discussed in Section 6.4.2. Each non-leaf node acts as a processor as well as a local control processor, passing and collecting information to and from its children.

Table 6.4.3 shows the main program in the control processor and Table 6.4.3 demonstrates how each slave processor acts in a parallel environment.

Their parallel algorithm performs well with some, not all topologies of a tree, because a parent node assigns different priority to each of its child node and transmits information according to the priorities of child nodes. Therefore, if a tree is very unbalanced, the 'lightest' subtree may finish all

---

[4] In the original description of this procedure, there were n classes, from C_1 to C_n. Since, class C_0 was used in the subsequence steps, we have modified this step to include $n+1$ classes, starting from C_0.

Table 6.9: Main program for the parallelisation of Dempster's combination rule

MAIN
1. decompose all subsets of Θ into n+1 classes
   $C_0, C_1, \ldots C_n$ evenly[4];
2. Pick up 2 mass functions and name them m_1 and m_2;
3. send pair (m_1, m_2) to each child node;
4. create an empty package Pak_0;
5. for each X in C_0, calculate $m(X)$ and append
   the result to package Pak_0;
6. receive package Pak_i from child i;
7. concatenate packages $Pak_0, \ldots, Pak_n$ to generate the
   combined mass $m_1 \oplus m_2$;
8. clear all packages;
9. if there is any unused mass function then
10. rename the combined mass function $m_1 \oplus m_2$ as m_1;
11. pick up one new mass function and name it as m_2;
12. goto step 3;
13. broadcast signal EOC (End Of Computation. Note: my explanation);
14. send the combined mass $m_1 \oplus \ldots \oplus m_p$ to each child;
15. create two empty packages Bel_0 and $Plau_0$ respectively;
16. for each X in C_0, calculate $Bel(X)$ and $Plau(X)$
    and append them to packages Bel_0 and $Plau_0$ respectively;
17. receive packages Bel_i and $Plau_i$ from child i, for all possible i;
18. concatenate Bel_i and $Plau_i$ separately to generate
    the belief and plausibility functions.

Table 6.10: Procedure for each individual processor

PE
1. get the node identity i;
2. while not detect signal EOC do;
3. receive pair (m_1, m_2) from its parent node;
4. if any child exists, send pair (m_1, m_2) to each child;
5. create an empty package Pak_i;
6. for each X in C_i, calculate $m(X)$ and append the result to package Pak_i;
7. if any child exists, receive package Pak_j from child j, for all possible j;
8. send all packages to its parent node;
9. receive the combined mass $m_1 \oplus \ldots \oplus m_p$ from it parent node;
10. if any child exists, send $m_1 \oplus \ldots \oplus m_p$ to each child;
11. create two empty packages Bel_i and $Plau_i$;
12. for each X in C_i, calculate $Bel(X)$ and $Plus(X)$ and append the result to packages Bel_i and $Plau_i$ respectively;
13. if any child exists, receive packages Bel_j and $Plau_j$ from child j, for all possible j;
14. send all packages to its parent node.

the task much earlier than the heavier ones and the heaviest may drag the whole processing down.

In [Maheshwari and Shen, 1998], a general purpose algorithm is proposed to partition an acyclic precedence graph (APG) into *clusters* such that all the clusters have almost balanced amount of computation load, and there is only one communication path between any pair of clusters. In this clustering algorithm, an APG is transformed first into a tree, then into a binary join tree, then into a balanced binary join tree, and finally into a collection of clusters partitioning the binary join tree. The available physical parallel environment determines the total number of clusters in the partition. By assigning each cluster to a processor, all the clusters can be dealt with simultaneously.

The idea of partitioning a binary join tree into clusters that was reported in [Maheshwari and Shen, 1998] has been adopted and extended in [Hong *et al.*, 1999] for combining multiple pieces of evidence based on Markov trees. On the basis of local computation technique, an efficient parallel algorithm is designed and evaluated in [Hong *et al.*, 1999] and [Hong, 2001] which involves the transformation of a Markov tree (Fig. 6.5) into a binary join tree (Fig. 6.6a) and partitioning the binary join tree into a list of clusters (Fig. 6.6b). Efficiency is achieved at the level of clusters with each cluster assigned to a single processor.

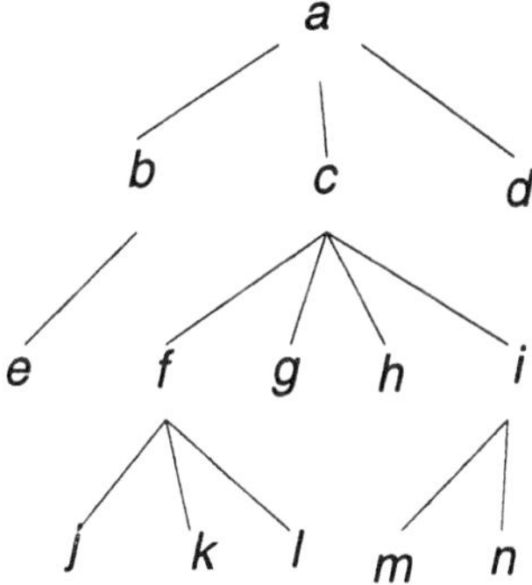

Figure 6.5: A Markov tree

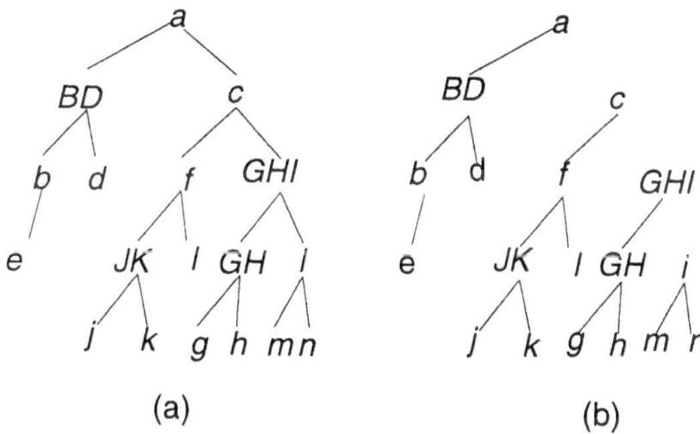

Here nodes in capital letters are auxiliary nodes added in when a Markov tree is transformed into a binary join tree

Figure 6.6: A binary join tree (a) and Four clusters formed from it (b)

6.5 Heuristic Knowledge Representation in DS Theory

Another limitation of DS theory is its weakness in representing heuristic knowledge when the theory is applied to expert systems. Much research has been done to extend the theory to tackle this difficulty, and we refer the reader to, e.g., [Lu and Stephanou, 1984], [Bonissone and Tong 1985], [Ginsberg, 1984], [Liu, 1986], [Yen, 1989], [Guan and Bell, 1991], and to [Liu et al, 1992] and [Liu *et al.*, 1994].

Considering the following piece of heuristic knowledge:

$$\text{if } X \textit{ is } X_1, \text{ then } Y \textit{ is } Y_1$$

with a degree of belief r_1, if we get a piece of evidence which says that X *is* X_1 with a degree of belief a_1, by invoking this rule, we should be able to obtain the corresponding degree y_1 for Y *is* Y_1. Certainly the value of y_1 should be dependent on both a_1 and r_1 (i.e., $y_1 = F(a_1, r_1)$, F being a function).

Generally, suppose a set of heuristic rules R includes:

R_1: if E_1 then H_{11} with a degree of belief r_{11};
H_{12} with a degree of belief r_{12};
...
R_2: if E_2 then H_{21} with a degree of belief r_{21};
H_{22} with a degree of belief r_{22};
...
...
R_n: if E_n then H_{n1} with a degree of belief r_{n1};
H_{n2} with a degree of belief r_{n2};
...

where $E_1, E_2, \ldots, E_n$ are values (or propositions) of variable E, and E_l is called an antecedent of rule R_l. H_{lj} in rule R_l is a subset of the values (or propositions) of variable H and it is called one of the conclusions of rule R_l. r_{lj} is called a rule strength.

Assume we have a piece of evidence which says that E_1 is confirmed with degree a_1, E_2 is confirmed with degree a_2, ..., E_n is confirmed with degree a_n, how can we solve the following problems:

1. What conditions should $\sum_l a_l$ satisfy?
2. What conditions should $\sum_j r_{lj}$ satisfy?
3. What is the function F to determine h_{lj}(the degree of belief on H_{lj}) from those a_l and r_{lj}?

4. If same conclusion H_{lj} is obtained from more than one set of rules, what will be the final degree of belief on H_{lj} from those r_{kt}?

5. What is the function F' to determine the degree of belief on the premise $(A_l \wedge\ B_j \wedge\ \ldots \wedge C_k)$, when variable E is the Cartesian product of variables $A, B, \ldots, C$, and each E_j is in a form of $(A_l \wedge\ B_j \wedge\ \ldots \wedge C_k)$, assuming evidence for $A, B, \ldots, C$, is known?

These problems have been modelled in fuzzy sets theory using a fuzzy extension of modal logic, based on Zadeh's concepts of necessity and possibility (cf. [Prade 1981]). They were also solved in MYCIN's certainty factor model (cf. [Shortliffe and Buchanan 1976]). Can these problems be solved in the Dempster-Shafer theory?

In [Yen, 1989], Yen used probabilistic mappings between two sets to replace multivalued mappings in DS theory in order to represent a heuristic rule as stated above.

Later, in [Liu et al, 1992], [Liu *et al.*, 1993], and [Liu *et al.*, 1994], this approach was extended to more general cases by using evidential mappings to replace probabilistic mappings. Advantages of this method and its comparison with Bayesian conditional probabilities [Pearl, 1988], with the approaches that were used in [Ginsberg, 1984] and [Hau and Kashyap, 1990], were fully discussed in [Liu *et al.*, 1994].

6.5.1 Yen's Probabilistic Mapping

Yen [Yen, 1989] proposed a probabilistic mapping to represent an uncertain supporting relationship between evidence and hypotheses. For a probability distribution on an evidence space (frame), a probabilistic mapping rather than a multipvalued mapping propagates it to the hypothesis space (frame). A mass function is then derived on the frame of hypothesis.

Namely, we have:

Definition 6.8: *Probabilistic Mapping* ([Yen, 1989]

A probabilistic multi-set mapping from a space E to a space Θ is a function Γ^*: $E \longrightarrow 2^{2^{\Theta_H} \times [0,1]}$.

The image of an element in E, denoted by $\Gamma^*(e_l)$, is a collection of subset-probability pairs, i.e.,

$$\Gamma^*(e_l) = \{(H_{l_1}, p(H_{l1})/e_l), \ldots, (H_{ln}, p(H_{ln})/e_l)\},$$

that satisfies the following conditions:

(1) $H_{lj} \neq \emptyset$,	$j = 1, \ldots, n$;
(2) $H_{lj} \cap H_{lj} = \emptyset$,	$j \neq k$;
(3) $p(H_{lj}/e_l) > 0$,	$j = 1, \ldots, n$;
(4) $\sum_j p(H_{lj}/e_l) = 1$.	

where e_l is an element of E , H_{l1}, ..., H_{ln} are subsets of Θ.

There are two limitations in Yen's extension:

1. Each evidential element in an evidence space is associated with a collection of non-empty disjoint hypothesis groups associated with their conditional probabilities.

   Therefore, it is difficult to represent ignorance such as, "*IF e is observed THEN the patient is likely to have a disease in $\{h_1, h_2\}$ with the probability 0.7, in $\{h_3, h_4\}$ with the probability 0.2, and in $\{h_1, h_2\}$, $\{h_3, h_4\}$ with probability 0.1*;

2. A belief distribution in an evidence space is measured by probabilities. If belief cannot be distributed on each individual element of an evidence space as probability, then the system fails to use this piece of evidence.

In [Liu et al, 1992] and [Liu *et al.*, 1994], an evidential mapping is proposed to replace Yen's probabilistic mapping.

6.5.2 Evidential Mapping

We start with:

Definition 6.9: *Evidential Mapping*

An evidential mapping is the mapping from one frame of discernment to another, which represents causal links among elements of two frames of discernment in the form of mass functions.

Formally, an evidential mapping from frame Θ_E to frame Θ_H is a function $\Gamma^*: \Theta_E \longrightarrow 2^{2^{\Theta_H} \times [0,1]}$.

The image of each element in Θ_E, denoted by $\Gamma^*(e_l)$, is a collection of subset-mass pairs, i.e.:

$$\Gamma^*(e_l) = \{(H_{l1}, f(e_l \to H_{l1})), \ldots, (H_{ln}, f(e_l \to H_{ln}))\},$$

that satisfies the following conditions:

a. $H_{lj} \neq \emptyset$,	$j = 1, \ldots, n;$
b. $f(e_l \to H_{lj}) > 0$,	$j = 1, \ldots, n;$
c. $\sum_j (e_l \to H_{lj}) = 1$.	

There is a set of heuristic rules, denoted as **R**, related to an evidential mapping, each of which is in the form of:

$$R_l : e_l \longrightarrow H_{l1\ (f(e_l \to H_{l1}))};$$
$$\ldots;$$
$$e_l \longrightarrow H_{ln\ (f(e_l \to H_{ln}))}.$$

where $f(e_l \to H_{lj})$ represents our belief exactly on H_{lj} given condition e_l, and it is in the range of [0,1].

A rule states that if e_l is true, then the truth of the problem carried by Θ_H is in H_{l1} with the degree of belief $f(e_l \to H_{l1})$ exactly committed to H_{l1}, ..., in H_{ln} with the degree of belief $f(e_l \to H_{ln})$ exactly committed to H_{ln}. e_l is called the antecedent of rule R_l and it is an element of Θ_E. H_{lj} is called one of the conclusions of rule R_l, and it is a subset of Θ_H. We name Θ_E and Θ_H as antecedent frame and conclusion frame of R respectively.

The corresponding matrix of an evidential mapping is:

row/col	H_1	H_2	$\ldots$	Θ
$\{e_1\}$	m_{11}	m_{12}	$\ldots$	m_{1s}
$\{e_2\}$	m_{21}	m_{22}	$\ldots$	m_{2s}
		$\ldots$		
$\{e_t\}$	m_{t1}	m_{t2}	$\ldots$	m_{ts}

$= M.$

The size of matrix M is $t \times s$ where t is the number of elements in Θ_E and s is equal to $|2^{\Theta_H}| - 1$ (except for $\emptyset$). H_l is a subset of the elements of Θ_H. For each H_{lj} appearing in $(H_{lj}, f(e_l \to H_{lj}))$ there is a column H_k where $H_k = H_{lj}$. The (l,k)-th entry of M is defined as m_{lk} which equals to $f(e_l \to H_{lj})$ if the pair $(H_{lj}, f(e_l \to H_{lj}))$ is an element of $\Gamma^*(e_l)$ and $H_k = H_{lj}$; otherwise m_{lk} is defined as zero. Thus, those $m_{l1}, m_{l2}, \ldots, m_{ls}$ of line l must satisfy the condition $\sum_j m_{lj} = 1$.

When a function m_l is defined as $m_l(A_{lj}) = m_{lj}$ for $j = 1, \ldots, s$, and $A_{l1} = \{(x,y) | x \in \neg\{e_l\} \text{ or } y \in H_{l1}\}, \ldots, A_{ln} = \{(x,y) | x \in \neg\{e_l\} \text{ or } y \in H_{ln}\}$, m_l is a mass function on $\Theta_E \times \Theta_H$, with its focal elements as A_{lj}.

Therefore, there are in total t mass functions on frame $\Theta_E \times \Theta_H$. But, we will show that the combination of any two of the above mass functions is meaningless.

In order to identify each row and column in M, we call H_k a *title of column k*, and $\{e_l\}$ – a *title of row i*. We also call $[\{e_1\}, \{e_2\}, \ldots, \{e_t\}]$ and $[H_1, H_2, \ldots, \Theta]$ a *row title vector*and a *column title vector* of M, respectively. When we mention matrix M of an evidential mapping, we assume the row title vector and the column title vector are known. Thus, for any given evidential mapping the related heuristic rule set and the matrix are unique.

An evidential mapping from Θ_E to Θ_H states that for two related questions represented by Θ_E and Θ_H, if the answer for the question represented by Θ_E is e_l, then the answer for the question represented by Θ_H is in a collection of subsets of Θ_H with e_l having different inter-relationships with different subsets in the collection. Value $f(e_l \to H_{lj})$ is used to reflect the sensitivity or strength of interrelationship between e_l and subset H_{lj}. Certainly, the total strength should be 1.

Example 6.5

If an evidential mapping Γ^* specifies a mapping from an antecedent space Θ_E to a conclusion space Θ_H as:

$$\begin{aligned}\Gamma^*(e_1) &= \{(\{a_1,a_2\},0.7),(\{a_3,a_4\},0.3)\}\\ \Gamma^*(e_2) &= \{(\{a_2,a_3\},0.8),(\Theta_H,0.2)\}\\ \Gamma^*(e_3) &= \{(\{a_4,a_5\},0.9),(\Theta_H,0.1)\}\end{aligned}$$

with a set of related heuristic rules given as:

R:

$$\begin{array}{ll} e_1 \longrightarrow \{a_1,a_2\}\ (0.7); & e_1 \longrightarrow \{a_3,a_4\}\ (0.3).\\ e_2 \longrightarrow \{a_2,a_3\}\ (0.8); & e_2 \longrightarrow \Theta_H\ (0.2).\\ e_3 \longrightarrow \{a_4,a_5\}\ (0.9); & e_3 \longrightarrow \Theta_H\ (0.1).\end{array}$$

where $\Theta_E = \{e_1,e_2,e_3\}$ and $\Theta_H = \{a_1,a_2,a_3,a_4,a_5\}$, then the corresponding matrix M has $2^5 - 1$ columns, most of which have only zero as values, such as columns $\{a_1\}, \{a_1,a_2,a_3\}$. Obviously, those columns with only zero values have no contribution to further inference and should be removed from a matrix in order to reduce the size of the matrix.

A matrix is called a *basic matrix*, denoted as *BM*, after removing all those columns which have only zero as values from a matrix.

Thus, the title vector of a basic matrix of an evidential mapping only contains those H_{lj} which appear in $\Gamma^*(e_l)$. The BM of the evidential mapping in the above example is:

row/col	$\{a_1,a_2\}$	$\{a_2,a_3\}$	$\{a_3,a_4\}$	$\{a_4,a_5\}$	Θ_H
$\{e_1\}$	0.7	0.0	0.3	0.0	0.0
$\{e_2\}$	0.0	0.8	0.0	0.0	0.2
$\{e_3\}$	0.0	0.0	0.0	0.9	0.1

$= BM$

with the row title vector being $[\{e_1\},\{e_2\},\{e_3\}]$ and the column title vector – $[\{a_1,a_2\}, \{a_2,a_3\}, \{a_3,a_4\}, \{a_4,a_5\}, \Theta_H\ ]$.

◇

Multivalued mappings in Definition 6.6 and Bayesian multi-valued causal link models ([Pearl, 1988]) can all be represented using evidential mappings.

Therefore, we have:

Corollary 3 *If all the m_{lj} in a basic matrix BM of an evidential mapping from Θ_E to Θ_H are either 1 or 0, then the evidential mapping is a multivalued mapping.*

For any e_l, the mass function m_l on $\Theta_E \times \Theta_H$ is a simple support function with a focal element $A_{lj}(A_{lj} = \{(x, y) | x \in \neg\{e_l\}$ or $y \in H_{lj}\})$, and $m_l(A_{lj}) = 1$.

Corollary 4 *If a basic matrix BM has $|\Theta_H|$ columns, and the titles of all columns are singletons of Θ_H, then the evidential mapping from Θ_E to Θ_H of this matrix is exactly a Bayesian multi-valued causal link model.*

For any e_l, the mass function m_l on $\Theta_E \times \Theta_H$ is a Bayesian probability distribution.

We refer to this kind of evidential mappings as Bayesian evidential mappings.

If a piece of evidence gives a probability distribution p on Θ_E, then a new function m on Θ_H can be calculated by the evidential mapping from Θ_E to Θ_H as:

$$m(H_k) = \begin{cases} \sum_l (p(e_l) m_{lk}) = \sum_l (p(e_l) f(e_l \to H_k)), \\ \qquad \textit{when } H_k \textit{ is the title of a column}, \\ 0, \ \textit{otherwise}. \end{cases} \tag{6.15}$$

Function m is a basic probability assignment on frame Θ_H and has the following features:

1. $m(\emptyset) = 0$,
2. $\sum_k m(H_k) = 1$, where $H_k \subseteq \Theta$.

This can be proved by the following procedure according to Definition 6.8, probability distribution p and features of a mass function:

$$\begin{aligned} \sum_k m(H_k) &= \sum_k \sum_l (p(e_l) f(e_l \to H_k)) \\ &= \sum_l \sum_k (p(e_l) f(e_l \to H_k)) \\ &= (\sum_l p(e_l))(\sum_k f(e_l \to H_k)) \\ &= (\sum_l p(e_l))(\sum_j f(e_l \to H_{lj})) \\ &\qquad \text{(because there exists an } H_{lj} \text{ such that } H_k = H_{lj}) \\ &= 1 \times 1 = 1. \end{aligned}$$

Hence:

Corollary 5 *Function m on frame Θ_H, derived from equation (6.15), is a mass function when p is a probability distribution on space Θ_E and Γ^* is an evidential mapping from Θ_E to Θ_H.*

Table 6.11: Result of combining two almost conflict information

suspect	witness1	witness2	combined result	unnormalized result
	m_1	m_2	m_{12}	m'_{12}
Peter	0.99	0.00	0.00	0.00
Paul	0.01	0.01	1.00	0.0001
Mary	0.00	0.99	0.00	0.00
$\emptyset$	0.00	0.00	0.00	0.9999

The theoretical support for equation (6.15) is the *law of total probability*:

$$p(A) = \sum_l p(A|B_l)p(B_l),$$

where B_l is an element of an exhaustive and mutually exclusive event set [Pearl, 1988].

We suppose that any evidence e giving $P(B_l/e)$ has no effect on $P(A|B_l)$. This rule is also called *Jeffrey's rule of conditioning* (see, for more details, [Jeffrey, 1965], [Shafer, 1981]).

In [Liu *et al.*, 1994], the techniques for *creating evidential mappings for incomplete heuristic rule sets*, for *predicting future events in DS theory*, and for *calculating posterior probabilities in DS theory* are fully discussed. Relationships between evidential mappings and Bayesian conditional probabilities are also studied in detail.

6.6 The Open World Assumption

DS theory is often criticized for combining two almost conflicting pieces of information (provided they are DS-independent). We can illustrate this problem with the following example:

Example 6.6 ([Smets, 1988])

Suppose we have a murder case with three suspects: Peter, Paul and Mary, and two witnesses. Table6.11 presents the degree of belief of each witness about who might be the murderer and the combined result of these two pieces of evidence.

◇

Zadeh [Zadeh, 1984] does not accept this solution, as it gives full certainty to a solution (Paul) that is hardly supported at all by the two witnesses. Looking at column 'Unnormalized Result' m'_{12}, this indicates that the 0.9999 portion of this belief has been committed to the empty set.

Smets [Smets, 1988] argued that in such a situation the meaning of the empty set should be reconsidered. Generally, when one considers a problem, one constructs three sets, the *Known as Possible (KP)* set including those propositions that are known to be possible, the *Unknown Proposition (UP)* set including those propositions for which one has no idea whether they are possible or impossible, and the *Known as Impossible (KI)* set including those propositions known as impossible. In DS theory, the *Unknown Proposition (UP)* set is always empty, and a frame of discernment is the *Known as Possible (KP)* set. In the above example, one solution to resolve the conflicting result is to accept the *Unknown Proposition* set and to believe that the real murderer must be a fourth person.

Another method for handling the present inconsistency is that a meta-level belief should be used to consider the reliability of the witnesses. Smets [Smets, 1988] addressed that certainly discounting is one way to take into account this meta-level belief. If a further piece of evidence says *The murderer is necessarily one of the group Peter, Paul and Mary,* we have to accept the *Unknown Proposition* set is empty and $m(\emptyset)$ should be 0. Therefore, $m(Peter) = m(Mary) = 0$ and $m(Paul) = 1$.

6.7 Summary

In this chapter, we first reviewed DS theory from the perspective of probability theory when considering the sensitive conditions on combination.

Then, we introduced techniques for solving computational complexity and for heuristic knowledge representation, as well as the open world assumption to the contribution of frame definition.

However, our major effort has been putting on clarifying the independent requirement in DS theory by defining the original information level and the target information level. We argue that any judgement of independence of evidence in DS theory should be made explicitly at the original information level.

An exclusive consideration of Dempster's combination rule without examining the original information will cause problems. However, Dempster's combination rule does not give us (or require from us) any information about what the original sources are.

So, the conclusion we get from the above analysis is that those counterintuitive examples given in some articles as, e.g., in [Black,1987], [Hunter,1987], [Lingras and Wong, 1990], [Nguyen and Smets, 1993], [Pearl, 1988], and in some of examples presented in [Voorbraak, 1991], show the weakness of the definition of frames.

The author argues that the accuracy of reasoning results depends on at which level the frame is constructed. Some other examples explain that even though Dempster's combination rule can be used in some situations, the results are still counterintuitive (violate common sense) like Example 6.8. This problem is also discussed in [Smets, 1988], [Zadeh, 1984] and [Zadeh, 1986]. Readers are referred to those papers if interested in more details.

Problems in Examples 2.4 and 3.3 are caused by ignoring the independent requirement defined by Dempster in his combination framework. In the sense of DS-independence required by Dempster's combination framework, these examples do not satisfy this requirement, so Dempster's combination framework is not applicable.

However, if we purely consider Dempster's combination rule and believe that those examples satisfy the independence requirement needed by Dempster's combination rule then Dempster's combination rule is applicable, with the combined results being counterintuitive.

From the former point of view, they are caused by the misapplication of the framework, from the latter point of view they are caused by the weakness of the combination formula. Neither of them is able to deal with such cases. Based on such a discussion, those belief functions, which can only be viewed as generalized probabilities, are precisely the cases which fail to satisfy the requirement of DS-independence. So, Dempster's combination rule is not suitable for coping with them.

Chapter 7

A Comprehensive Comparison of Generalized Incidence Calculus and Dempster–Shafer Theory

Generalized incidence calculus can be viewed either as a numerical or symbolic approach. It would be interesting to investigate its relationship with other pure numerical approaches, such as DS theory, when it is taken as a numerical approach.

In this chapter, DS theory is chosen as a representative of pure numerical mechanisms and a comprehensive comparison is provided between generalized incidence calculus and DS theory with respect to representing and combining evidence.

The comparison result shows that:

1. these two theories have the same ability in representing evidence,

2. any two pieces of evidence which can be combined using Dempster's combination rule can also be combined in incidence calculus by applying Proposition 5.1 and they give the same results, so that Dempster's combination rule and Proposition 5.1 are totally equivalent,

3. some dependent cases can only be dealt with by the new combination rule in generalized incidence calculus, not by Dempster's combination rule [Liu and Bundy, 1994].

Therefore, generalized incidence calculus may be taken as an alternative to DS theory to deal with some dependent evidence.

7.1 Comparison I: Representing Evidence

In DS theory, a mass function is defined on a frame Θ. However, given a set of atomic propositions P, P may not be a frame in its own right. Nevertheless, the basic elements set of P, $\mathcal{A}t$, is a frame.

In [Fagin and Halpern, 1989a], any arbitrarily defined frame Θ is taken to be a subset of some $\mathcal{A}t$ (in fact, given Θ, it is always possible to define $\mathcal{A}t$=Θ for a proper P). In this chapter, we follow the same idea and use $\mathcal{A}t$ to denote any frame of discernment.

Given a DS structure $(\mathcal{A}t, bel)$ and a GICT, $< \mathcal{W}, \mu, P, \mathcal{A}, i >$, where frame $\mathcal{A}t$ is defined to be the same as the basic element set of P, this DS structure is said to be *equivalent to the GICT* if for every $A \subseteq \mathcal{A}t$, there holds

$$bel(A) = Prob_*(\phi_A) = \mu(i_*(\phi_A))$$

.

Here, ϕ_A is defined as

$$\phi_A = \vee \delta_j$$

where $\delta_j \in A$.

In other words, $\mathcal{A}t$ is considered to be a frame when DS theory is used to describe degrees of belief of elements in $\mathcal{A}t$, and is treated as a collection of the basic elements formed from P when incidence calculus is employed to describe degrees of belief of elements in $\mathcal{A}t$.

A subset A of $\mathcal{A}t$ in $2^{\mathcal{A}t}$ is therefore semantically equivalent to a formula, $\vee\delta_j$ (where $\delta_j \in A$) in $\mathcal{L}(\mathcal{A}t)$, in the context that if formula $\vee\delta_j$ is true then the correct answer of the question represented by $\mathcal{A}t$ must be in subset A and vice versa.

Now, we have:

Theorem 15 *For any DS structure $(\mathcal{A}t, bel)$, there always exists an equivalent GICT $< \mathcal{W}, \mu, \mathcal{A}t, \mathcal{A}, i >$.*

Proof.

Given a DS structure $(\mathcal{A}t, bel)$, suppose that $A_{DS} = \{A_1, \ldots, A_n\}$ is the collection of all the focal elements, called the *focal element set*, of belief function bel, and m is the corresponding mass function, then $\Sigma m(A_j) = 1$.

Now:

1. Let a set of possible worlds $\mathcal{W}$ be $\{w_1, \ldots, xw_n\}$ and let $\mu(w_j) = m(A_j)$ for $j = 1, \ldots, n$. So, $\Sigma\mu(w_j) = 1$.
2. Let $\mathcal{A}'$ be a subset of $\mathcal{L}(P)$ as $\mathcal{A}' = \{\phi_{A_j} = \vee_l \delta_l \mid \delta_l \in A_j, A_j \in A_{DS}\}$.
3. Let function ii be $ii(\phi_{A_j}) = \{w_j\}$, $\phi_{A_j} \in \mathcal{A}'$; then $ii(\phi_{A_j}) \neq \emptyset$, $ii(\phi_{A_j}) \cap ii(\phi_{A_l}) = \emptyset$ and $\cup_{ii}(\phi_{A_j}) = \mathcal{W}$.

   Further, let $\mathcal{A}' = \mathcal{A}' \cup \{false\}$, and $ii(false) = \emptyset$.

   Therefore, ii is a basic incidence assignment on $\mathcal{A}'$ (cf. Definition 3.7).

4. Let $\mathcal{A}$ be the smallest set derived from $\mathcal{A}'$ based on Theorem 3 in Chapter 3. Define a function, i, from ii on $\mathcal{A}$ as follows: $i(\phi_A) = \cup_{\phi_{A_j} \in \mathcal{A}', \phi_{A_j} \models \phi_A} ii(\phi_{A_j})$, $\phi_{A_j} \in \mathcal{A}$; then, i is an incidence function.

Therefore, $< \mathcal{W}, \mu, \mathcal{A}t, \mathcal{A}, i >$ is a GICT.
For any formula ϕ_A in $\mathcal{L}(\mathcal{A}t)$ and its related subset A in $\mathcal{A}t$, we have:

$$\begin{aligned}
Prob_*(\phi_A) &= \mu(i_*(\phi_A)) \\
&= \mu(\bigcup_{\phi_{A_j} \models \phi_A} i(\phi_{A_j})) \\
&= \mu(\bigcup_{\phi_{A_j} \models \phi_A} (\bigcup_{\phi_{A_l} \models \phi_{A_j}} ii(\phi_{A_l}))) \\
&= \mu(\bigcup_{\phi_{A_l} \models \phi_A} ii(\phi_{A_l})) \\
&= \Sigma_{\phi_{A_l} \models \phi_A} \mu(ii(\phi_{A_l})) \\
&= \Sigma_{\phi_{A_l} \models \phi_A} \mu(\{w_l\}) \\
&= \Sigma_{A_l \subseteq A} m(A_l) \\
&= bel(A).
\end{aligned}$$

Belief function $bel(A)$ is therefore exactly the same as $Prob_*(\phi_A)$.

QED

As a consequence, $pls(A) = 1 - bel(\neg A) = 1 - p_*(\neg\phi_A) = \mu(\mathcal{W} \setminus i_*(\neg\phi_A)) = p^*(A)$.

This theorem states that a belief function on frame $\mathcal{A}t$ given by a DS structure has the same impact as the lower bound of probability distribution on formulae constructed from $\mathcal{A}t$ when $\mathcal{A}t$ is viewed as a basic element set.

Therefore, any belief function can be obtained as the lower bound of a probability distribution from a suitable GICT.

We use the following example to illustrate the procedure of generating a GICT from a DS structure:

Example 7.1 (re-visited Example 3.3)

> A person has four coats: two are blue and single-breasted, one is grey and double-breasted and one is grey and single-breasted. To choose which colour of coat this person is going to wear, one tosses a (fair) coin. Once the colour is chosen, to choose which specific coat to wear the person uses a mysterious nondeterministic procedure which we don't know anything about. What is the probability that the person is wearing a single-breasted coat?

DS structure:

Let a set of atomic propositions P be $P = \{g, d\}$ where g stands for "the coat is grey" and d stands for "the coat is double-breasted".

The corresponding basic element set is

$$\mathcal{A}t = \{g \wedge d, \neg g \wedge d, g \wedge \neg d, \neg g \wedge \neg d\},$$

which is a frame.

Element $\neg g \wedge d$ in this frame is false because there is no coat which is not grey but double-breasted. So, the original frame of discernment is reduced to

$$\mathcal{A}t = \{g \wedge d, g \wedge \neg d, \neg g \wedge \neg d\}.$$

One tossing a (fair) coin to decide which colour to choose indicates that a mass function on frame $\mathcal{A}t$ can be defined as

$$m(\{\neg g \wedge \neg d\}) = 0.5, \qquad m(\{g \wedge \neg d, g \wedge d\}) = 0.5,$$

with the focal element set A_{DS} as

$$A_{DS} = \{\{\neg g \wedge \neg d\}, \{g \wedge \neg d, g \wedge d\}\}.$$

Therefore, we obtained a DS structure $(\mathcal{A}t, bel)$. The degree of belief on $\neg d$ is $bel(\neg d) = m(\neg g \wedge \neg d) = 0.5$, and the degree of plausibility is 1.

The degrees of belief and plausibility say that the probability of the person wearing a single-breasted coat lies somewhere between 0.5 to 1 which cannot be measured by a single number.

Generalized incidence calculus:

According to Theorem 15 given above, the following steps will be taken to construct a GICT:

- Based on the DS structure given above, two possible worlds are defined: w_1 for blue and single-breasted coats and w_2 for grey coats; the probability of each of the possible worlds is 0.5.
- Let P and $\mathcal{A}t$ be the same as in DS structure, and $\mathcal{A}'$ be $\mathcal{A}' = \{\neg g \wedge \neg d, (g \wedge \neg d) \vee (g \wedge d)\}$.
- Define ii as $ii(\neg g \wedge \neg d) = w_1$, $ii((g \wedge \neg d) \vee (g \wedge d)) = w_2$, and $ii(false) = \emptyset$.
- Define an incidence function i as $i(\neg g \wedge \neg d) = w_1$, and $i((g \wedge \neg d) \vee (g \wedge d)) = w_2$.

Then, $< \mathcal{W}, \mu, P, \mathcal{A}, i >$ is a GICT.

From this GICT, it is easy to calculate i_* and i^*:

$$i_*(\neg d) = i(\neg g \wedge \neg d),$$

and

$$i^*(\neg d) = \mathcal{W} \setminus i_*(d) = \mathcal{W}.$$

Therefore,

$$Prob_*(\neg d) = 0.5, \qquad Prob^*(\neg d) = 1$$

which is identical to the result obtained from DS theory.

$\diamondsuit$

Next, we have:

Theorem 16 *For any GICT, $< \mathcal{W}, \mu, P, \mathcal{A}, i >$, there always exists an equivalent DS structure $(\mathcal{A}t, bel)$.*

Proof.

Suppose that $< \mathcal{W}, \mu, P, \mathcal{A}, i >$ is a GICT and ii is the corresponding basic incidence assignment, and:

1. Let the axioms in $\mathcal{A} \setminus \{false\}$ be $\phi_A, \phi_b, \ldots, \phi_C$.
2. Let a subset $\mathcal{A}_{DS}$ of $\mathcal{A}t$ be $A_{DS} = \{A \mid \phi_A \in \mathcal{A}\}$ (remember $\phi_A = \vee\delta_j, \delta_j \in A$).
3. Define $m(A_j) = \mu(ii(\phi_{A_j}))$ where $A_j \in A_{DS}$, then $\Sigma_{A_j} m(A_j) = 1$.

So, bel such that:

$$bel(A) = \Sigma_{B \subseteq A} m(B)$$

is a belief function on $\mathcal{A}t$, and $(\mathcal{A}t, bel)$ is a DS structure.

For any formula ϕ_A in $\mathcal{L}(\mathcal{A}t)$ and its related subset A of $\mathcal{A}t$, we have:

$$\begin{aligned} Prob_*(\phi_A) &= \mu(i_*(\phi_A)) \\ &= \mu(\bigcup ii(\phi_B) \mid \phi_B \models \phi_A, \phi_B \in \mathcal{A}) \\ &= \Sigma_{\phi_B}(\mu(ii(\phi_B)) \mid \phi_B \models \phi_A, \phi_B \in \mathcal{A}) \\ &= \Sigma(m(B) \mid B \subseteq A, B \in A_{DS}) \\ &= bel(A). \end{aligned}$$

Therefore, $< \mathcal{W}, \mu, P, \mathcal{A}, i >$ and $(\mathcal{A}t, bel)$ are equivalent.

QED

Example 7.2

This example, which is a continuation of Example 3.4, demonstrates the procedure of producing a DS structure based on a given GICT.

Assume we know that on *fri, sat, sun, mon* it will rain and on *mon, wed, fri* it will be windy. The question we are interested in is on which days it will not rain but be windy.

Generalized incidence calculus:

Let $\mathcal{W} = \{sun, mon, tues, wed, thus, fri, sat\}$ be the set of possible worlds, with a probability distribution $\mu(w_j) = 1/7$, for $j = 1, \ldots, 7$, and let $P = \{rainy, windy\}$.

An incidence function defined from the above description on a set of axioms $\mathcal{A}$=$\{rainy, windy$, $rainy \wedge windy$, $true$, $false\}$ is:

$$\begin{aligned} i(rainy) &= \{fri, sat, sun, mon\}, \\ i(windy) &= \{mon, wed, fri\}, \\ i(rainy \wedge windy) &= \{mon, fri\}, \\ i(true) &= \mathcal{W}, \\ i(false) &= \emptyset, \end{aligned}$$

and the basic incidence assignment ii is:

$$\begin{aligned} ii(rainy \wedge windy) &= \{fri, mon\}, \\ ii(rainy) &= \{sat, sun\}, \\ ii(windy) &= \{wed\}, \\ ii(false) &= \emptyset, \\ ii(true) &= ii(\mathcal{A}t) = \\ &= \{tues, thur\}. \end{aligned}$$

The corresponding basic element set $\mathcal{A}t$ is

$$\mathcal{A}t = \{rainy \wedge windy, rainy \wedge \neg windy, \neg rainy \wedge windy, \neg rainy \wedge \neg windy\}$$

and the GICT is

$$< \mathcal{W}, \mu, P, \mathcal{A}, i > .$$

From this GICT, we have:

$$\begin{aligned} i_*(\neg rainy \wedge windy) &= \emptyset, \\ i^*(\neg rainy \wedge windy) &= \mathcal{W} \setminus i_*(\neg(\neg rainy \wedge windy)) = \\ &= \{tues, wed, thus\}, \end{aligned}$$

so that

$$\begin{aligned} Prob_*(\neg rainy \wedge windy) &= 0, \\ Prob^*(\neg rainy \wedge windy) &= 3/7. \end{aligned}$$

DS structure:

On frame $\mathcal{A}t$ as defined above, a mass function m can be defined due to Theorem 16 as:

$$\begin{aligned} m(rainy \wedge windy) &= 2/7, \\ m(rainy) &= 2/7, \\ m(windy) &= 1/7, \\ m(\mathcal{A}t) &= 2/7. \end{aligned}$$

Moreover, $bel(\neg rainy \wedge windy) = 0$ and $pls(\neg rainy \wedge windy) = 3/7$. DS structure $(\mathcal{A}t, bel)$ gives therefore the same result as that given VIA generalized incidence calculus.

The result obtained says that it is very possible that it will not rain but be windy on Tuesday, Wednesday or Thursday.

The above two theorems together prove the equivalence of DS theory and generalized incidence calculus in terms of representing uncertain information. However, they attain this purpose in rather different ways.

Bundy in [Bundy, 1992] summarized this difference as follows:

> "... both systems[1] permit only partial definition of the probabilities of some formulae. DS theory achieves this by defining the incidence of all formulae, but not defining the probabilities of all the possible worlds, i.e., Γ is a total function, but μ' is a partial function. Incidence calculus achieves a similar effect the other way round, i.e., μ is total but i is partial".

A similar result was also obtained in [Correa da Silva and Bundy, 1990]. In their paper, it is proved that any original incidence calculus theory is equivalent to a total Dempster-Shafer probability structure, and any total Dempster-Shafer probability structure is equivalent to an original incidence calculus theory. In this chapter, we have shown that GICTs are equivalent to DS structures.

7.2 Comparison II: Combining DS-Independent Evidence

For any two DS structures $(\mathcal{A}t, bel_1)$ and $(\mathcal{A}t, bel_2)$, if these two belief functions are derived from two DS-independent pieces of evidence, then these two

[1]DS theory and incidence calculus.

belief functions can be combined using Dempster's combination rule.

On the other hand, from these two DS structures, two GICTs can also be produced, and their combination leads to the third GICT using Proposition 5.1. What we need to prove in such a situation is that the combination result of the two DS structures should be equivalent to the combined GICT.

We have the following theorem:

Theorem 17 *Suppose* $(\mathcal{A}t, bel_1)$ *and* $(\mathcal{A}t, bel_2)$ *are two DS structures and* bel_1 *and* bel_2 *are obtained from two DS-independent pieces of evidence and assume that the combined DS structure is* $(\mathcal{A}t, bel)$.

Further let $< \mathcal{W}_1, \mu_1, \mathcal{A}t, \mathcal{A}_1, i_1 >$ *and* $< \mathcal{W}_2, \mu_2, \mathcal{A}t, \mathcal{A}_2, i_2 >$ *be the two GICTs produced from* $(\mathcal{A}t, bel_1)$ *and* $(\mathcal{A}t, bel_2)$, *and* $< \mathcal{W}, \mu, \mathcal{A}t, \mathcal{A}, i >$ *be the combined GICT.*

Then, $(\mathcal{A}t, bel)$ *is equivalent to* $< \mathcal{W}, \mu, \mathcal{A}t, \mathcal{A}, i >$. *That is, for any subset* A *of* $\mathcal{A}t$, *there is*

$$bel(A) = Prob_*(\phi_A).$$

Our proof is divided into two parts. The first part proves that the conflict weight k in the combined DS structure is equal to $\mu(\mathcal{W}_0)$ in the combined GICT, and the second part two proves that $bel(A) = p_*(\phi_A)$, for any $A \subseteq \mathcal{A}t$.

Because bel_1 and bel_2 are derived from two DS-independent pieces of evidence, two probability spaces $(\mathcal{W}_1, 2^{\mathcal{W}_1}, \mu_1)$ and $(\mathcal{W}_2, 2^{\mathcal{W}_2}, \mu_2)$ should be DS-independent. So, Proposition 5.1 is applicable to combining these two derived GICTs.

Proof.

Let the two focal element sets in these two DS structures $(\mathcal{A}t, bel_1)$ and $(\mathcal{A}t, bel_2)$ be:

$$\begin{array}{lcll} A_{DS} & = & \{A_1, A_2, \ldots, A_n\}, & \Sigma m_1(A_l) = 1, \\ B_{DS} & = & \{B_1, B_2, \ldots, B_m\}, & \Sigma m_2(B_j) = 1. \end{array}$$

The combined DS structure is $(\mathcal{A}t, bel)$ with bel defined as $bel_1 \oplus bel_2$. Furthermore, let the two sets of axioms in the corresponding two GICTs:

$$< \mathcal{W}_1, \mu_1, P, \mathcal{A}_1, i_1 > \text{ and } < \mathcal{W}_2, \mu_2, P, \mathcal{A}_2, i_2 >$$

be:

$$\mathcal{A}_1 = \{\phi_{A_1}, \phi_{A_2}, \ldots, \phi_{A_n}\}, ii_1(\phi_{A_l}) = \{w_{1l}\}, \mu_1(w_{1l}) = m_1(A_l),$$

and

$$\mathcal{A}_2 = \{\psi_{B_1}, \psi_{B_2}, \ldots, \psi_{B_m}\}, ii_2(\phi_{B_j}) = \{w_{2j}\}, \mu_2(w_{2j}) = m_2(B_j).$$

Part I: Proof of $k = \mu(\mathcal{W}_0)$, where k is the weight of the conflict between these two DS structures, and $\mathcal{W}_0$ – which is defined in Chapter 5 – is the conflict set in the combined GICT.

Suppose that $m = m_1 \oplus m_2$. For any $A_l \cap B_j = \emptyset$ ($A_l \in A_{DS}, B_j \in B_{DS}$), $m_1(A_l)m_2(B_j)$ will be part of k. That is, $k = k' + m_1(A_l)m_2(B_j)$.

Equivalently, for $\phi_{Al} \in \mathcal{A}_1$ and $\psi_{B_j} \in \mathcal{A}_2$, $\phi_{A_l} \wedge \psi_{B_j} \models false$ should be held.

So:

$$\begin{aligned}
\mu(\mathcal{W}_0) &= \mu(\textstyle\bigcup_{\phi_{A_l}\wedge\psi_{B_j}\models false} i_1(\phi_{A_l}) \otimes i_2(\psi_{B_j}))\\
&= \mu(\textstyle\bigcup_{\phi_{A_l}\wedge\psi_{B_j}\models false}(\bigcup_{\phi_{A'_l}\models\phi_{A_l}} ii_1(\phi_{A'_l})) \otimes (\bigcup_{\psi_{B'_j}\models\psi_{B_j}} ii_2(\psi_{B'_j})))\\
&= \mu(\textstyle\bigcup_{\phi_{A_l}\wedge\psi_{B_j}\models false}(\bigcup_{\phi_{A'_l}\wedge\psi_{B'_j}\models\phi_{A_l}\wedge\psi_{B_j}} ii_1(\phi_{A'_l}) \otimes ii_2(\psi_{B'_j})))\\
&= \mu(\textstyle\bigcup_{\phi_{A'_l}\wedge\psi_{B'_j}\models false} ii_1(\phi_{A'_l}) \otimes ii_2(\psi_{B'_j}))\\
&= \Sigma(\mu_1(ii_1(\phi_{A'_l}))\mu_2(ii_2(\psi_{B'_j})) \mid \phi_{A'_l} \wedge \psi_{B'_j} \models false)\\
&= \Sigma(\mu_1(w_{1l'})\mu_2(w_{2j'}) \mid \phi_{A'_l} \wedge \psi_{B'_j} \models false)\\
&= \Sigma(m_1(A'_l)m_2(B'_j) \mid A'_l \cap B'_j = \emptyset)\\
&= k.
\end{aligned}$$

Part II: Proof of $bel(A) = p_*(\phi_A)$, for any $A \subseteq \mathcal{A}t$.

For any subset C of $\mathcal{A}t$, and its corresponding formula φ_C, we need to prove that $bel(C) = Prob_*(\varphi_C)$.

For $A_l \in A_{DS}$ and $B_j \in B_{DS}$, if $A_l \cap B_j \subseteq C$, then $m_1(A_i)m_2(B_j)$ is part of $bel(C)$.

Equivalently, for ϕ_{A_l} in $\mathcal{A}_1$ and ψ_{B_j} in $\mathcal{A}_2$, we have $\phi_{A_l} \wedge \psi_{B_j} \models \varphi_C$.

So:

$$\begin{aligned}
&Prob_*(\varphi_C) =\\
&= \mu(i_*(\varphi_C))\\
&= \mu(\bigcup_{\phi_{A_l}\wedge\psi_{B_j}\models\varphi_C} i_1(\phi_{A_l}) \otimes i_2(\psi_{B_j}))\\
&= \Sigma_{\phi_{A_l}\wedge\psi_{B_j}\models\varphi_C}(\mu_1(i_1(\phi_{A_l}))\mu_2(i_2(\psi_{B_j})))/(1-k)\\
&= \Sigma_{\phi_{A_l}\wedge\psi_{B_j}\models\varphi_C}(\mu_1(\bigcup_{\phi_{A'_l}\models\phi_{A_l}} ii_1(\phi_{A'_l}))\mu_2(\bigcup_{\psi_{B'_j}\models\psi_{B_j}} ii_2(\psi_{B'_j})))/(1-k)\\
&= \Sigma_{\phi_{A_l}\wedge\psi_{B_j}\models\varphi_C}(\mu_1(\bigcup_{\phi_{A'_l}\models\phi_{A_l}} \{w_{1l'}\})\mu_2(\bigcup_{\psi_{B'_j}\models\psi_{B_j}} \{w_{2j'}\}))/(1-k)\\
&= \Sigma_{\phi_{A'_l}\wedge\psi_{B'_j}=\varphi_{C'},\varphi_{C'}\models\varphi_C}(\mu_1(\{w_{1l'}\})\mu_2(\{w_{2j'}\}))/(1-k)
\end{aligned}$$

Table 7.1: Combination of two DS-independent pieces of evidence

A	$\{a,b,c\}$	$\mathcal{A}t$
m	0.7	0.3
$\{c,d\}$	$\{c\}$	$\{c,d\}$
0.6	0.42	0.18
$\mathcal{A}t$	$\{a,b,c\}$	$\mathcal{A}t$
0.4	0.28	0.12

$$\begin{aligned} &= \Sigma_{C' \subseteq C, A'_l \cap B'_j = C'}(m_1(A'_l)m_2(B'_j))/(1-k) \\ &= \Sigma_{C' \subseteq C} m(C') \\ &= bel(C). \end{aligned}$$

QED

Example 7.3

This example demonstrates the procedure of combining two DS-independent pieces of evidence using both Dempster's combination rule and Proposition 5.1 in generalized incidence calculus.

Using Dempster's combination rule:

Assume two DS structures:

$$(\mathcal{A}t, bel_1) \quad \text{and} \quad (\mathcal{A}t, bel_2)$$

are known with the following additional information:

$$\begin{aligned} &\mathcal{A}t = \{a,b,c,d\}, \\ &A_{DS} = \{\{a,b,c\}, \mathcal{A}t\}, \\ &B_{DS} = \{\{c,d\}, \mathcal{A}t\}, \\ &m_1(\{a,b,c\}) = 0.7, m_1(\mathcal{A}t) = 0.3, \\ &m_2(\{c,d\}) = 0.6, m_2(\mathcal{A}t) = 0.4. \end{aligned}$$

Here, A_{DS} and B_{DS} are the focal element sets of belief functions bel_1 and bel_2, respectively.

Combining these two belief functions derived from m_1 and m_2, a joint belief function is derived as shown in Table 7.1.

From this table, it is possible to calculate degrees of belief on any subsets of $\mathcal{A}t$. For instance, for subset $\{a,b,c\}$, its degree of belief is $bel(\{a,b,c\}) = 0.42 + 0.28 = 0.7$ and its degree of plausibility is $pls(\{a,b,c\}) = 1$.

Table 7.2: Combination of two DS-independent GICTs

We ignored axiom false in both GICTs.

ϕ_A $i(\phi_A)$	$a \vee b \vee c$ $\{w_{11}\}$	*true* $\mathcal{W}_1$
$c \vee d$ $\{w_{21}\}$	c $\{w_{11}\} \otimes \{w_{21}\}$ 0.42	$c \vee d$ $\mathcal{W}_1 \otimes \{w_{21}\}$ 0.6
true $\mathcal{W}_2$	$a \vee b \vee c$ $\{w_{11}\} \otimes \mathcal{W}_2$ 0.7	*true* $\mathcal{W}_1 \otimes \mathcal{W}_2$ 1

Using the incidence calculus combination rule:

From the two DS structures given above, we are able to form two GICTs as:

$$< \mathcal{W}_1, \mu_1, P, \mathcal{A}_1, i_1 >,$$

and

$$< \mathcal{W}_2, \mu_2, P, \mathcal{A}_2, i_2 >,$$

with the following additional information:

$$\begin{array}{lll}
\mathcal{A}_1 = \{a \vee b \vee c, \mathcal{A}t, false\}, & \mathcal{A}_2 = \{c \vee d, \mathcal{A}t, false\}, & P = \mathcal{A}t, \\
\mathcal{W}_1 = \{w_{11}, w_{12}\}, & \mu_1(w_{11}) = 0.7, & \mu_1(w_{12}) = 0.3, \\
\mathcal{W}_2 = \{w_{21}, w_{22}\}, & \mu_2(w_{21}) = 0.6, & \mu_2(w_{22}) = 0.4, \\
i_1(a \vee b \vee c) = \{w_{11}\}, & i_1(\mathcal{A}t) = \mathcal{W}_1, & i_1(false) = \emptyset; \\
i_2(c \vee d) - \{w_{21}\}, & i_2(\mathcal{A}t) = \mathcal{W}_2, & i_2(false) = \emptyset.
\end{array}$$

As $(\mathcal{W}_1, 2^{\mathcal{W}_1}, \mu_1)$ and $(\mathcal{W}_2, 2^{\mathcal{W}_2}, \mu_2)$ are DS-independent, Proposition 5.1 is used to combine these two GICTs as given in Table 7.2.

The combined GICT is

$$< \mathcal{W}_1 \otimes \mathcal{W}_2, \mu, P, \mathcal{A}, i >$$

where $\mu(< w_{1l}, w_{2j} >) = \mu_1(w_{1l})\mu_2(w_{2j})$.

From this theory, we are also able to calculate the degree of our belief in any formula. For example, $Prob_*(a \vee b \vee c) = \mu(i_*(a \vee b \vee c)) = 0.7$ and $Prob^*(a \vee b \vee c) = 1$ which are the same as what is obtained in DS theory.

Comparing Table 7.1 and Table 7.2, we will find that these two structures give the same result (numerically) on any subset (or formula). We will also find that whenever a numerical value (mass value) appears in Table 7.1, a corresponding incidence set replaces its position in Table 7.2.

Table 7.3: All the possible labels, the number of balls with each possible label, and the subset name containg balls with same label

Set of labels X	Number of balls having that label	Subset in $\mathcal{W}$
axy	4	W_1
ax	4	W_2
ay	16	W_3
a	16	W_4
bxy	10	W_5
bx	10	W_6
by	20	W_7
b	20	W_8

The combination procedure in generalized incidence calculus combines possible worlds instead of numbers. The degree of belief in a formula is calculated based on the incidence set.

Combining evidence on sets in generalized incidence calculus suggests that it may be possible to cancel overlaps implied in two pieces of information.

In the next section we are going to show that indeed some dependent cases which Dempster's combination has failed to deal with can be coped with in generalized incidence calculus.

7.3 Comparison III: Combining Dependent Evidence

In this section, we first examine an example which can be dealt with using the combination rule in generalized incidence calculus but cannot be dealt with using Dempster's combination rule.

We then explore the theoretical difference between these two theories to explain why DS theory fails to deal with dependent evidence while incidence calculus succeeds.

Example 7.4 (re-visited Example 6.4)

Suppose that there are 100 labelled balls in an urn as shown in Table 7.3. Each ball must have either label a or b, but not both, in addition to some extra labels, x, y or xy.

Agents A and B give separate observations of drawing a ball from the urn as follows:

Agent A: The drawn ball has label x. The set of labels describing those balls is $X_1 = \{axy, ax, bxy, bx\}$ with the probability distribution:

$$\mu_1(\{axy\}) = \mu_1(\{ax\}) = 4/28,$$

and

$$\mu_1(\{bxy\}) = \mu_1(\{bx\}) = 10/28.$$

Probability $\mu_1(\{axy\}) = 4/28$ means that a drawing ball has label axy with probability $4/28$ when it is known that the label definitely contains x.

Agent B: The drawn ball has label y. The set of labels describing those balls is $X_2 = \{axy, ay, bxy, by\}$ with the probability distribution:

$$\mu_2(\{axy\}) = 4/50, \quad \mu_2(\{ay\}) = 16/50,$$

and

$$\mu_2(\{bxy\}) = 10/50 \quad \mu_2(\{by\}) = 20/50.$$

Probability $\mu_2(\{axy\}) = 4/50$ means that a drawing ball has label axy with probability $4/50$ when it is known that the label definitely contains y.

Based on these two pieces of evidence, we are interested in knowing the degree of our belief that the drawn ball also has label b.

Using Dempster's combination rule:

Let $\{a, b\}$ be a frame of discernment, where a stands for 'the drawn ball has label a' and b stands for 'the drawn ball has label b'.

Two mass functions are defined on S based on the information carried by two agents A and B as:

$$\begin{aligned} m_A(a) &= 2/7 \\ m_A(b) &= 5/7, \end{aligned}$$

$$\begin{aligned} m_B(a) &= 2/5, \\ m_B(b) &= 3/5, \end{aligned}$$

where $m_A(a)$ is the mass value on a given by agent A's observation which represents the possibility of a ball having label a when the ball is observed having label x, and $m_B(a)$ is the mass value on a given by agent B's observation which represents the possibility of a ball having label a when the ball is observed having label y.

The result of applying Dempster's combination rule to m_X and m_Y is $m(b) = m_X \oplus m_Y(b) = 15/19$, so that $bel(b) = 15/19$, which is believed to be wrong.

Using the incidence calculus combination rule:

Now we study this example in generalized incidence calculus.

Let a set of possible worlds $\mathcal{W}$ contain 100 labelled balls, and

$$\mathcal{W} = W_1 \cup W_2 \cup W_3 \cup W_4 \cup W_5 \cup W_6 \cup W_7 \cup W_8,$$

where W_1 contains 4 possible worlds in which a drawing ball has label axy, ..., W_8 contains 20 possible worlds in which a drawing ball has label b.

The probability distribution on $\mathcal{W}$ is $\mu(w) = 1/100$ for any $w \in \mathcal{W}$.

Furthermore, let a set of atomic propositions P be $\{a, b, x, y\}$ where a means that the chosen ball has label a, etc.

From observations made by agents A and B, it is possible to construct two GICTs

$$< \mathcal{W}, \mu, P, \mathcal{A}_1, i_1 >,$$

and

$$< \mathcal{W}, \mu, P, \mathcal{A}_2, i_2 >,$$

where:

$$\begin{array}{l}
i_1(x) = W_1 \cup W_2 \cup W_5 \cup W_6, \\
i_1(a \wedge x) = W_1 \cup W_2, \quad i_1(b \wedge x) = W_5 \cup W_6, \\
i_1(a \wedge x \wedge y) = W_1, \quad i_1(b \wedge x \wedge y) = W_5, \\
i_1(true) = \mathcal{W}, \quad i_1(false) = \emptyset, \\
i_2(y) = W_1 \cup W_3 \cup W_5 \cup W_7, \\
i_2(a \wedge y) = W_1 \cup W_3, \quad i_2(b \wedge y) = W_5 \cup W_7, \\
i_2(a \wedge x \wedge y) = W_1, \quad i_2(b \wedge x \wedge y) = W_5, \\
i_2(true) = \mathcal{W}, \quad i_2(false) = \emptyset,
\end{array}$$

in which:

$$\mathcal{A}_1 = \{x, a \wedge x, b \wedge x, a \wedge x \wedge y, b \wedge x \wedge y, true, false\}$$

and

$$\mathcal{A}_2 = \{y, a \wedge y, b \wedge y, a \wedge x \wedge y, b \wedge x \wedge y, true, false\}$$

.

Applying the combination rule given in Chapter 5 to these two GICTs, we obtain the third GICT, $< \mathcal{W}, \mu, P, \mathcal{A}, i >$, with

$$\begin{aligned}
\mathcal{A} &= \{x \wedge y, a \wedge y, b \wedge y, a \wedge x, b \wedge x, x, y, a \wedge x \wedge y, \\
&\qquad b \wedge x \wedge y, a \wedge b \wedge x \wedge y, true, false\} \\
i(a \wedge b \wedge x \wedge y) &= \emptyset, \\
i(b \wedge x \wedge y) &= W_5, \\
i(a \wedge x \wedge y) &= W_1, \\
i(x \wedge y) &= W_1 \cup W_5, \\
i(a \wedge x) &= W_1 \cup W_2, \\
i(a \wedge y) &= W_1 \cup W_3, \\
i(b \wedge x) &= W_5 \cup W_6, \\
i(b \wedge y) &= W_5 \cup W_7, \\
i(x) &= W_1 \cup W_2 \cup W_5 \cup W_6, \\
i(y) &= W_1 \cup W_3 \cup W_5 \cup W_7, \\
i(true) &= \mathcal{W}, \\
i(false) &= \emptyset.
\end{aligned}$$

It is easy to prove that, for any $\phi \in \mathcal{A}$, $i_*(\phi) = i^*(\phi) = i(\phi)$, so function *Prob*, defined as $Prob(\phi) = Prob_*(\phi) = \mu(i(\phi))$, is a probability distribution on $\mathcal{A}$. In particular, $Prob(b \wedge x \wedge y) = 10/100$ and $Prob(x \wedge y) = 14/100$.

Therefore, by equation (3.4) from Chapter 3, we have

$$Prob(b \mid x \wedge y) = \frac{Prob(b \wedge x \wedge y)}{Prob(x \wedge y)} = \frac{\mu(i(b \wedge x \wedge y))}{\mu(i(x \wedge y))} = 5/7.$$

This result is consistent with what we could get via probability theory.

◇

Now we try to explain theoretically why Dempster's combination rule cannot be used in this case.

In fact, the two mass functions are derived from two probability spaces $(S_1, 2^{S_1}, \mu_1)$ and $(S_2, 2^{S_2}, \mu_2)$, where $S_1 = W_1 \cup W_2 \cup W_5 \cup W_6$ with $\mu_1(s) = 1/28$ and $S_2 = W_1 \cup W_3 \cup W_5 \cup W_7$ with $\mu_2(s) = 1/50$. These two probability spaces are defined from the unique space $(\mathcal{W}, 2^{\mathcal{W}}, \mu)$ and they share the information carried by the subset $W_1 \cup W_5$.

Because DS theory is a pure numerical uncertainty reasoning mechanism, it is not possible to precisely and explicily represent the joint (or overlapped) part of the information provided by two pieces of evidence. Therefore Dempster's combination rule is not applicable.

Table 7.4: Combination of two DS-independent GICTs

We ignored axiom false in both GICTs. Let $S_1 = W_1 \cup W_2 \cup W_5 \cup W_6$, $S_2 = W_1 \cup W_3 \cup W_5 \cup W_7$, *and let* $\phi = a \wedge b \wedge x \wedge y$

ϕ_A $i(\phi_A)$	$a \wedge x$ $W_1 \cup W_2$	$b \wedge x$ $W_5 \cup W_6$	x S_1	$a \wedge x \wedge y$ W_1	$b \wedge x \wedge y$ W_5	*true* $\mathcal{W}$
$a \wedge y$ $W_1 \cup W_3$	$a \wedge x \wedge y$ W_1	ϕ $\emptyset$	$a \wedge x \wedge y$ W_1	$a \wedge x \wedge y$ W_1	ϕ $\emptyset$	$a \wedge y$ $W_1 \cup W_3$
$b \wedge y$ $W_5 \cup W_7$	ϕ $\emptyset$	$b \wedge x \wedge y$ W_5	$b \wedge x \wedge y$ W_5	ϕ $\emptyset$	$b \wedge x \wedge y$ W_5	$b \wedge y$ $W_5 \cup W_7$
y S_2	$a \wedge x \wedge y$ W_1	$b \wedge x \wedge y$ W_5	$x \wedge y$ $W_1 \cup W_5$	$a \wedge x \wedge y$ W_1	$b \wedge x \wedge y$ W_5	y S_1
$a \wedge x \wedge y$ W_1	$a \wedge x \wedge y$ W_1	ϕ $\emptyset$	$a \wedge x \wedge y$ W_1	$a \wedge x \wedge y$ W_1	ϕ $\emptyset$	$a \wedge x \wedge y$ W_1
$b \wedge x \wedge y$ W_5	ϕ $\emptyset$	$b \wedge x \wedge y$ W_5	$b \wedge x \wedge y$ W_5	ϕ $\emptyset$	$b \wedge x \wedge y$ W_5	$b \wedge x \wedge y$ W_5
true $\mathcal{W}$	$a \wedge x$ $W_1 \cup W_2$	$b \wedge x$ $W_5 \cup W_6$	x S_2	$a \wedge x \wedge y$ W_1	$b \wedge x \wedge y$ W_5	*true* $\mathcal{W}$

In generalized incidence calculus, instead of combining numerical values on set $\mathcal{A}t$, we combine two pieces of evidence symbolically at the original information level, i.e., at the probability space level.

In the above example, since the two probability spaces are somehow related to the unique space $(\mathcal{W}, 2^{\mathcal{W}}, \mu)$, we establish two generalized incidence functions from $\mathcal{W}$ to P rather than from S_1 and S_2 to P, respectively. Therefore, it is possible to cancel the overlapped information carried by the two observations.

Hence, we conclude that even though the two theories have the same ability in representing evidence and combining DS-independent information, their theoretical structures are rather different.

The essence of incidence calculus, *indirect encoding of probabilities of formulae,* makes it possible to cancel the effect of overlapped information and to provide an alternative combination mechanism which combines dependent information.

Although an attempt to combine dependent information at the probability space level was considered in [Shafer, 1982] and in [Lingras and Wong, 1990], no unique rule was provided in DS theory for general cases because of the theoretical limitation of the theory.

7.4 Comparison IV: Some Other Aspects of the Two Theories

7.4.1 Recovering Mass Functions

In DS theory, when a belief function *bel* is known, its mass function can be recovered from it when the frame $\mathcal{A}t$ is finite. This is particularly necessary when Dempster's combination rule is applied because this rule only uses mass functions.

Since a belief function (*bel*) on a frame $\mathcal{A}t$ is equivalent to the lower bound ($Prob_*$) of a probability distribution on $\mathcal{A}t$, the application of Algorithm B given in Chapter 4 on a DS structure $(\mathcal{A}t, bel)$ would certainly be able to recover its corresponding mass function (cf. [Liu *et al.*, 1993a]). This is given in Algorithm C shown below.

Algorithm C: Deriving Mass Functions:

Input: *$Prob_*$ on set $\mathcal{A} = \mathcal{L}(P)$*
Output: **if** *$Prob_*$ is a belief function on $\mathcal{A}$* **then** *recover its mass function on this frame*[2]

[2] In fact, this language set can be any frame of discernment.

Step 1: for each *formula* $\phi \in \mathcal{A}$ **do**

if $Prob_*(\phi) = 0$ **then** $\mathcal{A} \leftarrow \mathcal{A} \setminus \{\phi\}$

choose *a subset* $\mathcal{A}_0$ *of* $\mathcal{A} \setminus \{false\}$ *as* $\mathcal{A}_0 = \{\psi_1, \ldots, \psi_n\}$ **such that:**

for each $\psi_j \in \mathcal{A}_0$ **do**

for each $\phi \in \mathcal{A}$ **do**

if $\phi \neq \psi_j$ **then** $\phi \not\models \psi_j$

for each $\psi_j \in \mathcal{A}_0$ **do**

$m(\psi_j) \leftarrow Prob_*(\psi_j)$,

$Prob'_*(\psi_j) \leftarrow Prob_*(\psi_j)$

$\mathcal{A}' \leftarrow \mathcal{A} \setminus \mathcal{A}_0$

Step 2: choose *a formula* ψ *in* $\mathcal{A}'$ **such that**

for each $\psi' \in \mathcal{A}'$ **do**

if $\psi \neq \psi'$ **then** $\psi' \not\models \psi$

$Prob'_*(\psi) \leftarrow Prob_*(\psi) - \Sigma_{\phi_j \in \mathcal{A}_0, \phi_j \models \psi} Prob'_*(\phi_j)$

if $Prob'_*(\psi) > 0$ **then**

$\mathcal{A}_0 \leftarrow \mathcal{A}_0 \cup \{\psi\}$,
$\mathcal{A}' \leftarrow \mathcal{A}' \setminus \{\psi\}$,
$m(\psi) \leftarrow Prob'_*(\psi)$

else if $Prob'_*(\psi) = 0$ **then** $\mathcal{A}' \leftarrow \mathcal{A}' \setminus \{\psi\}$ *(ψ is not a focal element of this belief function)*

else if $Prob'_*(\phi) < 0$ **then** *this assignment is not a belief function, stop the procedure*

Step 3: if $\mathcal{A}' \neq \emptyset$ **then go to Step 2**

end *of algorithm*

Finally, all the elements in $\mathcal{A}_0$ will be the focal elements of this belief function and function m defined in Step 2 is the corresponding mass function. It is easy to prove that $\Sigma_A m(A) = 1$ and $m(\emptyset) = 0$.

This algorithm tries to find the focal elements of a belief function one by one. Once all the focal elements are found and the mass values they carry are calculated, the corresponding mass function is known.

Example 7.5

Assume there are four elements in $\mathcal{A}t = \{a, b, c, d\}$, and $\mathcal{A} = \mathcal{L}(P)$ is:

$$\begin{aligned} \mathcal{A} &= \\ &= \{a, b, c, d, a \vee b, a \vee c, a \vee d, b \vee c, b \vee d, c \vee d, \\ &\quad a \vee b \vee c, a \vee c \vee d, a \vee b \vee d, b \vee c \vee d, \\ &\quad a \vee b \vee c \vee d = true, false\}. \end{aligned}$$

Let the corresponding numerical assignment $Prob_*$ on $\mathcal{A}$ be:

$$\begin{array}{ll} Prob_*(\{a\}) = 0.5, & Prob_*(\{d\}) = 0.3, \\ Prob_*(\{a \vee b\}) = 0.7, & Prob_*(\{a \vee c\}) = 0.5, \\ Prob_*(\{a \vee d\}) = 0.8, & Prob_*(\{b \vee d\}) = 0.3, \\ Prob_*(\{c \vee d\}) = 0.3, & Prob_*(\{a \vee b \vee c\}) = 0.7, \\ Prob_*(\{a \vee c \vee d\}) = 0.8, & Prob_*(\{a \vee b \vee d\}) = 1, \\ Prob_*(\{b \vee c \vee d\}) = 0.3, & Prob_*(\{a \vee b \vee c \vee d\} = true) = 1, \end{array}$$

with the lower bounds of all other formulae being equal to.

Applying Algorithm C on $(\mathcal{A}, Prob_*)$, the calculation procedure for a mass function is as follows:

Step 1. After deleting those elements with 0 degree of belief, we have:

$\mathcal{A} = \{a, d, a \vee c, a \vee b, a \vee d, b \vee d, c \vee d, a \vee b \vee c, a \vee c \vee d, a \vee b \vee d, b \vee c \vee d, a \vee b \vee c \vee d = true\}$.

$\mathcal{A}_0 = \{a, d\}$.

$m(a) = Prob'_*(a) = Prob_*(a) = 0.5$,

$m(d) = Prob'_*(d) = Prob_*(d) = 0.3$.

$\mathcal{A}' = \mathcal{A} \setminus \mathcal{A}_0$.

Step 2. Choose $a \vee c$ from $\mathcal{A}'$.

Because $a \models a \vee c$, then we have $Prob'_*(a \vee c) = Prob_*(a \vee c) - Prob'_*(a) = 0.5 - 0.5 = 0$. $a \vee c$ is not a focal element. So, define $\mathcal{A}' = \mathcal{A}' \setminus \{a \vee c\}$.

Repeat this procedure until we get $a \vee b$ and we have $Prob'_*(a \vee b) = 0.7 - 0.5 = 0.2$.

Define:

$$\begin{aligned} m(a \vee b) &= Prob'_*(a \vee b) = 0.2, \\ \mathcal{A}_0 &= \mathcal{A}_0 \cup \{a \vee b\}, \\ \mathcal{A}' &= \mathcal{A}' \setminus \{a \vee b\}. \end{aligned}$$

Step 3. Repeat Step 2 until $\mathcal{A}'$ is empty.

Eventually, $\mathcal{A}_0$ is constructed as $\{a, d, a \vee b\}$ and the mass function m gives $m(a) = 0.5, m(d) = 0.3, m(a \vee b) = 0.2$.

$\diamondsuit$

7.4.2 Recovering Probability Spaces

In [Fagin and Halpern, 1989a] and [Fagin and Halpern, 1989b] a method of assigning probability measures on formulae instead of sets was suggested. In their approach, given a probability space $(\mathcal{W}, \chi, \mu)$, an inner measure on a propositional language set $\mathcal{L}(P)$ can be defined through a truth assignment $\pi(w) : \mathcal{L}(P) \rightarrow \{\textbf{true}, \textbf{false}\}$ as follows.

If $\pi(w)(\phi) = \textbf{true}$, then ϕ is said to be true at w; otherwise ϕ is false at w. ϕ^π is defined to contain all those elements in $\mathcal{W}$ in which ϕ is true.

If we define $Prob_*(\phi) = \mu_*(\phi^\pi)$ where μ_* is the inner measure of μ, then $Prob_*$ is called an inner measure of a probability distribution on $\mathcal{L}(P)$.

It is proved in [Fagin and Halpern, 1989b] that a belief function on such a language set is also an inner measure generated from a probability space.

Therefore it is also interesting to apply the above technique to recover a probability space when an inner measure $Prob_*$ of probabilities on $\mathcal{L}(P)$ is known.

More details about probability spaces and their relationship with DS theory can be found in [Fagin and Halpern, 1989a], [Fagin and Halpern, 1989b].

Algorithm D: Recovering Probability Spaces:

input: *inner measure* $Prob_*$ *(or bel) on* $\mathcal{A} = \mathcal{L}(P)$
output: $(\mathcal{W}, \chi, \mu)$ *and* π *from which* $Prob_*$ *(or bel) is derived.*

Step 1: for each *formula* $\phi \in \mathcal{A}$ **do**

  if $Prob_*(\phi) = 0$ **then** $\mathcal{A} \leftarrow \mathcal{A} \setminus \{\phi\}$

 choose *a subset* $\mathcal{A}_0$ *of* $\mathcal{A} \setminus \{false\}$ *as* $\mathcal{A}_0 = \{\psi_1, \ldots, \psi_n\}$ **such that:**

  for each $\psi_j \in \mathcal{A}_0$ **do**

   for each $\phi \in \mathcal{A}$ **do**

    if $\phi \neq \psi_j$ **then** $\phi \not\models \psi_j$

 $\chi' \leftarrow \emptyset$

 for each $\phi_j \in \mathcal{A}_0$ **do**

  $\chi' \leftarrow \chi' \cup \{w_j\}$,

$\mu(w_j) = Prob_*(\phi_j),$

$\pi(w_j)(\phi_j) \leftarrow$ **true**,

$Prob'_*(\phi_j) \leftarrow Prob_*(\phi_j)$

$\mathcal{A}' \leftarrow \mathcal{A} \setminus \mathcal{A}_0$

$l \leftarrow \mid \mathcal{A}_0 \mid$

Step 2: choose *a formula* ψ *in* $\mathcal{A}'$ **such that:**

for each $\psi' \in \mathcal{A}'$ **if** $\psi \neq \psi'$ **then** $\psi' \not\models \psi$

$Prob'_*(\psi) \leftarrow Prob_*(\psi) - \Sigma_{\phi_j \in \mathcal{A}_0, \phi_j \models \psi} Prob'_*(\phi_j)$

if $Prob'_*(\psi) > 0$ **then**

$\mathcal{A}_0 \leftarrow \mathcal{A}_0 \cup \{\psi\},$
$\mathcal{A}' \leftarrow \mathcal{A}' \setminus \{\psi\},$
$\mu(w_{l+1}) \leftarrow Prob'_*(\psi),$
$\pi(w_{l+1})(\psi) \leftarrow$ **true**,
$l \leftarrow l + 1$

else if $Prob'_*(\psi) = 0$ **then** *there is no need to create an extra possible world to match* ψ

else if $Prob'_*(\phi) < 0$ **then** *this assignment is not a correct inner measure, stop the procedure*

Step 3: if $\mathcal{A}' \neq \emptyset$ **then go to Step 2**

end *of algorithm*

Set $\chi' = \{\{w_1\}, \ldots, \{w_{l+1}\}\}$ is the basis of an unspecified probability space. It is easy to prove that $\Sigma_{w_j} \mu(w_j) = 1$.

The corresponding probability space will be $(\mathcal{W}, \chi, \mu)$ where χ is the $\sigma-$algebra generated by the basis χ'. In the simplest case, the probability space can just be $(\chi', 2^{\chi'}, \mu)$.

According to Fagin and Halpern's method describe above, it can be proved that the lower bound $Prob_*$ on $\mathcal{L}(P)$ can be re-calculated from probability space $(\chi', 2^{\chi'}, \mu)$ and mapping function π.

Example 7.6

Continuing Example 7.5, if we take $Prob_*$ as an inner measure of a probability measure on $\mathcal{A}$ from an unknown probability space, this space can be recovered as $(\mathcal{W}, \chi, \mu)$, where the basis for χ is $\chi' = \{\{w_1\}, \{w_2\}, \{w_3\}\}$, and $\mu(w_1) = 0.5, \mu(w_2) = 0.3, \mu(w_3) = 0.2$.

The corresponding π assigns **true** to axioms in $\mathcal{A}_0$ at the following possible worlds, and **false** at any other possible worlds.

So:

$$\begin{aligned}\pi(w_1)(a) &= \textbf{true},\\ \pi(w_2)(d) &= \textbf{true},\\ \pi(w_3)(a \vee b) &= \textbf{true}.\end{aligned}$$

◇

7.5 Summary

In this chapter, we have made a comprehensive comparison between DS theory and generalized incidence calculus in the following three aspects:

1. representation of evidence,
2. combining DS-independent evidence, and
3. dealing with dependent evidence.

We conclude that these two theories have the same ability in representing incomplete information and combining DS-independent evidence. However, generalized incidence calculus is superior to DS theory in coping with overlapped information.

DS theory is a pure numerical approach while generalized incidence calculus possesses both symbolic and numerical features. That is, generalized incidence calculus can make an inference either at the symbolic level by producing incidence sets or on the numerical basis by calculating lower or upper bounds on probabilities of formulae.

The new combination rule in generalized incidence calculus is proposed based on the symbolic feature of the theory.

Trying to combine dependent pieces of information using Dempster's combination rule has been investigated in [Dubois and Prade,1986], [Smets, 1990], [Nguyen and Smets, 1993], [Kennes, 1991], [Shafer, 1986], [Shafer, 1987], [Lingras and Wong, 1990]. [Dubois and Prade,1986] intended to model this problem in terms of set theory in fuzzy logic.

In the works [Smets, 1990] and [Nguyen and Smets, 1993], the authors discussed the possible methods of combining dependent information in the transferable belief model. A possibility of solving this problem in the concept of a *category* is proposed in [Kennes, 1991].

In [Lingras and Wong, 1990], [Shafer, 1986] and [Shafer, 1987] approaches were presented which are closer to ours in generalized incidence calculus.

In [Shafer, 1986] and [Shafer, 1987], Shafer demonstrated, using examples, that some dependent evidence can be combined using Dempster's combination framework, but cannot be combined using Dempster's combination rule.

In [Lingras and Wong, 1990] a compatibility view is adopted when combining two pieces of evidence. According to their method of compatibility view, two pieces of evidence are combined at the original information level before being propagated to the target space. The combination relies on the compatibility relation among two spaces which, they assume, is defined by the user.

The joint probability distribution is defined based on whether the two spaces are probabilistically independent. Two original evidence spaces are assumed to be probabilistically independent if the spaces are logically independent. By logically independent, they mean that if every element in X_1 is compatible with all elements in X_2 then X_1 and X_2 are said to be logically independent. If two spaces are logically independent, they can be dealt with using Dempster's rule. Otherwise, either a Bayesian approach or a dependency function is used to get a joint probability on the target space. However the definition of logically independent on two spaces is not sufficient to guarantee that the two probability distributions are independent as we have seen in Example 6.3 given by Shafer.

In summary, all these papers we referred to above tried to either re-explain Dempster's combination rule in alternative terminologies, or complement the current Dempster's combination rule from different perspectives. None of them tried to give a new combination mechanism which keeps the spirit of Dempster's combination idea but is distinct from it.

In contrast to the approaches shown above, we proposed a new rule and tried to combine several pieces of evidence at the original information level. In this new combination rule, original sources are required in order to carry out the combination, so it is natural and convenient to make the independent judgement using the evidence sources.

Dempster's combination rule is a special case of this new rule when several sources are DS-independent. The main advantage of the new rule is to unify the combination and propagation procedures into one structure using generalized incidence functions.

In general, independent relations among multiple sources of evidence can be considered as special cases of dependent situations. As Pearl indicated (cf. [Pearl, 1992]):

> "...if we have several items of evidence, each depending on the state of nature, these items of evidence should also depend on each other. This kind of dependency is not a nuisance but a necessary bliss; no evidential reasoning would otherwise be possible".

In our combination rule, we have indeed adopted the same idea and made some efforts towards combining dependent evidence. This result would be useful for further research work on either this topic or relevant topics. It tells us that it is a promising way to cancel the overlapped and duplicated information from several pieces of evidence at the symbolic level rather than at the numerical level.

Chapter 8

Assumption-Based Truth Maintenance Systems

The aim of this chapter is to provide a basis for the discussion in Chapter 9. We will not review all the work on the assumption-based truth maintenance system (ATMS) carried out so far. Rather, we will only focus on two aspects of the ATMS:

- its basic reasoning mechanism, and
- the possibility of associating it with numerical uncertainty mechanisms.

After we have made these two aspects clear, we are then ready to walk through the next chapter which shows the relationship between the ATMS and generalized incidence calculus.

8.1 Reasoning Mechanism in the ATMS

8.1.1 Structure of Nodes

The truth maintenance system (TMS) ([Doyle, 1979]), and later the ATMS ([de Kleer, 1986a] and [de Kleer, 1986b]), are both symbolic approaches to recording and maintaining dependencies among statements or beliefs.

The central issue in such a system is that for a derived statement, a set of more primitive ones supporting this statement will be produced.

Each primitive statement, which is called an *assumption*, denoted by a capital letter in an ATMS, is assumed to be true if there is no conflict. Assumptions can be interpreted as the primitive data from which all other statements (or beliefs or data) can be derived.

Each statement, no matter it is an assumption or a derived statement, is represented as a unique node in an ATMS. A group of assumptions ensuring the truth of a statement is called an *environment* of this statement.

The collection of all such groups is known as the *label* of this statement. Therefore, the label of a statement contains all possible situations in which this statement is true.

Formally, an ATMS node is stated in the following form:

$$node_{statement} :< statement, label, justifications >,$$

where $node_x$ designates the node with datum (statement) x.

In order to generate the label for a statement, knowledge about supporting relationships among statements is crucial. In an ATMS, a direct supporting relationship between a particular statement and several other statements is represented by a *justification*. In other words, a justification of a statement (or node) identifies a group of other statements (or nodes) from which the current statement can be directly derived. Justifications are specified by the system designer.

For instance, if we have three inference rules like:

$$r_1 : if\ \ it\ \ rains \rightarrow grass\ \ is\ \ wet,$$

$$r_2 : if\ \ a\ \ sprinkler\ \ is\ \ on \rightarrow grass\ \ is\ \ wet,$$

$$r_3 : if\ \ grass\ \ is\ \ wet \rightarrow cannot\ \ mow\ \ grass,$$

then logically we can infer that

$$r_4 : if\ \ it\ \ rains \rightarrow cannot\ \ mow\ \ grass$$

or

$$r_5 : if\ \ a\ \ sprinkler\ \ is\ \ on \rightarrow cannot\ \ mow\ \ grass.$$

In an ATMS, if r_1, r_2, r_3, r_4 and r_5 are represented by $node_1$, $node_2$, $node_3$, $node_4$ and $node_5$, respectively, then $node_4$ is derivable from the conjunction of $node_1$ and $node_3$, and $node_5$ is derivable from the conjunction of $node_2$ and $node_3$. We say that $(node_1, node_3)$ is a justification of $node_4$ and $(node_2, node_3)$ is a justification of $node_5$.

To take this example a bit further by assuming that r_1, r_2, and r_3 are valid under the condition that assumptions A, B and C are true, respectively (such as, A stands for 'the rain is reasonably heavy - enough to get things wet', B for 'the sprinkler works fine and has been on for a while with water coming out', and C for 'my lawn mower will get damaged if I use it when the grass is wet'), there is no doubt then that rule r_4 is valid only when $A \wedge C$ is true, denoted as $\{A, C\}$.

Technically speaking, $\{A\}, \{B\}$, $\{C\}$ and $\{A, C\}$, $\{B, C\}$ are environments of r_1, r_2, r_3, r_4, and r_5, respectively.

For instance, rule r_4, known also as $node_4$ here, can be represented by the following structure, if we at the same time do not exclude other possible situations of deriving it:

$$node_4 :< q_1 \rightarrow q_4, \{\{A, C\} \ldots\}, \{(node_1, node_3) \ldots\} > .$$

Here, q_1 and q_4 stand for 'it rains' and 'cannot cut grass', respectively. We also let q_2 stand for 'a sprinkler is on' and q_3 for 'grass is wet' as well.

In summary, an ATMS structure requires the following items of messages:

node: A node (called a problem-solver's datum) in an ATMS representing any datum unit used in the system. This datum unit can be a fact, a proposition or any formula in a propositional language which the system uses.

assumptions: A set of distinguished nodes which are believed to be true without requiring any precondition.

justifications: A justification of a node specifying those nodes from which it can be derived. A node with several justifications suggests multiple routes of inferring the node. Justifications are supplied by the problem-solver.

label: The label of a node consisting of all the possible environments. Each environment is a collection of assumptions under which the statement represented by the node is interpreted as true.

nogood: A special node in an ATMS system whose label consists of all environments in which falsity can be derived.

8.1.2 Types of Nodes

All nodes in an ATMS are divided into four categories:

- *assumed nodes*,
- *assumption nodes*,
- *premises*, and
- *derived nodes.*

r_1, r_2 and r_3 are true under the support of assumptions A, B and C respectively. So, $node_1$, $node_2$ and $node_3$ below are named *assumed nodes* and are represented as:

$$node_1 :< q_1 \rightarrow q_3, \{\{A\}\}, \{(A)\} >,$$

$$node_2 :< q_2 \rightarrow q_3, \{\{B\}\}, \{(B)\} >,$$

Table 8.1: A set of ATMS nodes, their types, labels and justifications

Node	Node Type	Statement	Label	Justification
N_1	assumed node	$q_1 \to q_3$	$\{\{A\}\}$	$\{(A)\}$
N_2	assumed node	$q_2 \to q_3$	$\{\{B\}\}$	$\{(B)\}$
N_3	assumed node	$q_3 \to q_4$	$\{\{C\}\}$	$\{(C)\}$
N_4	derived node	$q_1 \to q_4$	$\{\{A, C\}\}$	$\{(N_1, N_3)\}$
N_5	derived node	$q_2 \to q_4$	$\{\{B, C\}\}$	$\{(N_2, N_3)\}$
N_6	assumption node	A	$\{\{A\}\}$	$\{(A)\}$
N_7	assumption node	B	$\{\{B\}\}$	$\{(B)\}$
N_8	assumption node	C	$\{\{C\}\}$	$\{(C)\}$
N_9	premise node	q_1	$\{\{\}\}$	$\{()\}$
N_{10}	derived node	q_3	$\{\{A\}\}$	$\{(N_1, N_9)\}$
N_{11}	derived node	q_4	$\{\{A, C\}\}$	$\{(N_1, N_3, N_9), (N_4, N_9)\}$

$$node_2 :< q_3 \to q_4, \{\{C\}\}, \{(C)\} > .$$

A, B, and C known as assumptions, are also represented using nodes These nodes are called *assumption nodes* and are represented as:

$$node_6 :< A, \{\{A\}\}, \{(A)\} >,$$
$$node_7 :< B, \{\{B\}\}, \{(B)\} >,$$
$$node_8 :< C, \{\{C\}\}, \{(C)\} > .$$

$Node_4$ is a *derived node*, since it is derived from $node_1$ and $node_3$.

Apart from the above three types of nodes, there are some facts (statements) which may hold universally. A *premise* is a node representing such a fact or statement. A premise, such as q_1, is in the form

$$node_9 :< q_1, \{\{\}\}, \{()\} > .$$

In summary, each assumption node takes itself as its justification and label. On the other hand, an assumed node contains only assumption(s) both in its justification and in its label. However, a derived node takes other nodes other than assumptions in its justifications. A premise does not require any justification and assumption to support its true.

Table 8.1 summarises the nodes that have been created so far and that will be created if q_1 is also known.

Although the ATMS has been praised and applied in many areas, such as, diagnostic and abductive reasoning, because of its ability in recording dependencies among statements, it has also been criticised for its limitation in representing statements which are probably true rather than absolutely true.

In order to overcome this limitation, the integration of a numerical method with an ATMS has been intensively investigated in the last decade, exemplified by [d'Ambrosio, 1988], [d'Ambrosio, 1990], [de Kleer and Williams, 1987], [Dubois *et al.*, 1990], [Fulvio Monai and Chehire, 1992], [Pearl, 1988], [Laskey and Lehner, 1989], [Provan, 1989] and [Liu and Bundy, 1996].

For example, in the original ATMS structure, a rule like, $q \rightarrow r$ is true with probability 0.5 cannot be represented. However, if this rule is re-written as $q \wedge A \rightarrow r$ where A is an assumption under which the rule is true and A is true with probability 0.5, then it is easy to associate a probability with the rule. Assumption A, also called an auxiliary node, is introduced to carry the uncertainty.

Once suitable assumptions have been created to carry uncertainties, an ATMS system will be able to calculate the degree of belief of a node through the uncertainties attached to assumptions in its label set. In this way, any heuristic rule $a \rightarrow b$ with rule strength m can be changed as (cf. [Pearl, 1988]):

$$a \wedge C \rightarrow b.$$

In this expression, C stands for how strong an agent believes in b if a is known to be true.

8.2 Non-Redundant Justification Sets and Environments

To summarise the main concept in the previous section, we conclude that there are four types of nodes in an ATMS. Some of these nodes have labels specified initially, like assumed, premise and assumption nodes. Derived nodes rely on their justifications to obtain their labels.

As we have seen above, an ATMS node is stated in the following form:

$$node_{statement} :< statement, label, justifications > .$$

Both the label and the justifications of a node can be explained using material implication. Given a node c with label $\{\{A_1, A_2, \ldots\}\ \{B_1, B_2, \ldots\} \ldots\}$ and with justifications $\{(z_1,\ z_2,\ \ldots)\ (y_1,\ y_2,\ \ldots) \ldots\}$, the conjunction of the assumptions in each environment of its label makes c true, such as $A_1 \wedge A_2 \ldots$ of environment $\{A_1, A_2 \ldots\}$.

So, $L(c)$ is the collection of those elements each of which makes c true, $L(c) = \{(A_1 \wedge A_2 \wedge \ldots), (B_1 \wedge B_2 \wedge \ldots), \ldots\}$.

Alternatively, it is possible to write

$$(A_1 \wedge A_2 \wedge \ldots) \vee (B_1 \wedge B_2 \wedge \ldots) \vee \ldots \rightarrow c.$$

Similarly, j being a justification of node c states that the conjunction of the elements in j logically supports node c. If $j = (z_1, z_2, \ldots)$ is a justification of c, then $\wedge_l z_l$ is a formula in a propositional language which implies c, that is, $\wedge_l z_l \models c$.

In general if we let:

$$J(c) = \{(z_1 \wedge z_2 \wedge \ldots), (y_1 \wedge y_2 \wedge \ldots), \ldots\},$$

then every element in $J(c)$ semantically implies c, denoted as $J(c) \models c$.

Therefore, there is a similar implication relation:

$$(z_1 \wedge z_2 \wedge \ldots) \vee (y_1 \wedge y_2 \wedge \ldots) \vee \ldots \rightarrow c.$$

Also, given a well defined ATMS, it is possible to infer the label for any derived node through justifications as follows: if the justification set of node r_n contains l justifications as

$$\{(r_{11}, \ldots, r_{t1}), \ldots, (r_{1l}, \ldots, r_{t'l})\},$$

then the label set of r_n is

$$L(r_n) = \cup_{k=1}^{l}(L(r_{1k}) \otimes \ldots \otimes L(r_{jk})),$$

when $L(r_{11})$,..., $L(r_{t'l})$ are known.

Each environment in a label consists of a set of non-redundant assumptions. That is, the deleting of any element in an environment will destroy the implication relation between this environment and its node. Also, for any two environments of one node, they surely do not imply each other. Otherwise, one environment will be a proper subset of another, and the latter will certainly be redundant.

The same rules also apply to the justifications of a node. It is assumed in an ATMS that every justification is non-redundant and any two justifications do not imply each other.

If we require that a justification set of a node is non-redundant, then deleting any justification from the justification set of a node will result in losing a path from which the node can derive.

This issue can well be illustrated in the following example taken from [Laskey and Lehner, 1989].

Example 8.1

Assume there are following five inference rules:

$r_1 : e \rightarrow d,$ $\qquad r_2 : d \rightarrow b,$

$r_3 : b \rightarrow a,$ $\qquad r_4 : d \rightarrow c,$

$r_5 : c \rightarrow a,$

with five assumptions Z, X, V, Y and W supporting them, respectively.

These five rules can be encoded into a set of ATMS assumed nodes as follows:

$$\begin{array}{ll} node_1: & < e \rightarrow d, \{\{Z\}\}, \{(Z)\} >, \\ node_2: & < d \rightarrow b, \{\{X\}\}, \{(X)\} >, \\ node_3: & < b \rightarrow a, \{\{V\}\}, \{(V)\} >, \\ node_4: & < d \rightarrow c, \{\{Y\}\}, \{(Y)\} >, \\ node_5: & < c \rightarrow a, \{\{W\}\}, \{(W)\} > . \end{array}$$

From this five initial rules, rules $d \rightarrow a$ and $e \rightarrow a$ can be derived, and be denoted by the following two derived nodes:

$$\begin{array}{ll} node_6: & < d \rightarrow a, \{\{X, V\}, \{Y, W\}\}, \{(node_2, node_3), (node_4, node_5)\} >, \\ node_7: & < e \rightarrow a, \{\{Z, X, V\}, \{Z, Y, W\}\}, \{(node_1, node_6)\} >, \end{array}$$

If $node_6$ is replaced by its justifications whenever it appears in the justification set of $node_7$, then $node_7$ is changed as:

$$\begin{array}{ll} node_7: & < e \rightarrow a, \{\{Z, X, V\}, \{Z, Y, W\}\}, \\ & \{(node_1, node_2, node_3), (node_1, node_4, node_5)\} > . \end{array}$$

It is worth noting that the labels of $node_6$ and $node_7$ are empty before we make any inference using their justifications (remember justifications are supplied by the designer and they are not empty).

We should also notice that if we add another node (for instance, $node_4$) to the set $(node_1, node_2, node_3)$, the extended justification set also implies $node_7$, but we do not extend any justifications of $node_7$ in this way, because of the requirement of non-redundant justifications.

In fact, there are in total seven conjunctions of nodes making $node_7$ true, but only two of them are included in the justification set. These two justifications are the non-redundant ones which are essential to identify all the possible inference routes for $node_7$.

8.3 Probabilistic Assumption Sets

In an ATMS, all nodes can be divided into four types:

- *assumptions,*
- *assumed nodes,*
- *premises* , and
- *derived nodes.*

If we follow the convention that assumptions cannot become another type of node and non-assumptions cannot become assumptions [de Kleer, 1986a], then it is possible to keep all assumption nodes in one set and the rest of nodes in another, and these two sets are disjoint.

An agent's belief in a statement represented by a node can have one of the three possible values:

- believed,
- disbelieved, and
- unknown.

Given a node c, if one of the environments in its label is believed, then c is believed. If one of the environments in the label of $\neg c$ is believed, then c is disbelieved, otherwise c is unknown. A conflict is detected when both c and $\neg c$ are believed.

This suggests that there exit two sets of environments, the conjunction of which would support falsity. In other words, the *nogood* node has an non-empty label.

Therefore, some of the assumptions included in the two environments for c and $\neg c$ should be revised to be false in order to solve this conflict. As we have discussed earlier, this kind of inference in an ATMS cannot produce results with numerical degrees of belief.

Attempts of attaching numerical uncertainty values to assumptions in the ATMS enable an ATMS to calculate degrees of belief in statements.

The belief of a node is defined as the probability of its label, $Bel(c) = \mu(L(c))$.

One common requirement in all these extensions of the ATMS is that probabilities assigned to assumptions must be assumed probabilistically independent in order to calculate the degree of belief in a statement[1].

For example, in ([Pearl, 1988]), the rule:

turn the key $\rightarrow$ *start the engine with* 0.8

can be represented in an ATMS as $node_{b \rightarrow a}$ given as:

$$node_{b \rightarrow a} :< b \rightarrow a, \{\{B\}\}, \{(B)\} >,$$

[1]Except for [Dubois *et al.*, 1990] and [Fulvio Monai and Chehire, 1992] in which the topic was not discussed. Also in [Liu and Bundy, 1996], this requirement is relaxed if a common probability space from which several dependent probability spaces are constructed is known.

where B stands for an assumption (or a set of assumptions) which supports the implication relation $b \rightarrow a$ and it is assigned with probability value 0.8; a and b represent propositions *'start the engine'* and *'turn the key'*, respectively.

Assume that assumed node, $node_b$, given below is also in the ATMS system,

$$node_b :< b, \{\{A\}\}, \{(A)\} >,$$

then $node_a$ can be created for statement a as

$$node_a :< a, \{\{A, B\}\}, \{(node_{b \rightarrow a}, node_b)\} > .$$

a is a derived node, and its label is $\{\{A, B\}\}$.

Therefore:

$$Bel(a) = \mu(L(a)) = \mu(A \wedge B) = \mu_A(A) \times \mu_B(B) = 0.8,$$

if we assume that the probability distributions are probabilistically independent and the action *'turn the key'* is true, i.e., $\mu_A(A) = 1$.

Once probabilities are assigned on assumptions then, in principle, an ATMS has the ability to make plausible inferences with beliefs.

For a simple case like above, the calculation of probabilities on nodes is not difficult to perform. However, in most cases labels of nodes are very complicated and probability distributions on assumptions may be somehow related. In those circumstances, the calculating of probabilities of labels of nodes is a complex task as shown in [Laskey and Lehner, 1989] and [Pearl, 1988].

Here, in this book, we give the following two definitions to cope with this difficulty in general. The motivation of proposing the following two definitions is stimulated by the idea of managing possible worlds in incidence calculus. These two definitions together provide a theoretical basis for associating and managing probabilities in an ATMS. It covers the related work in [Laskey and Lehner, 1989] and [Pearl, 1988].

Definition 8.1: *Probabilistic assumption set*[2]

A set $\{A_1, \ldots, A_n\}$, denoted as $S_{A_1,\ldots,A_n}$, is called a probabilistic assumption set for assumptions $A_1, \ldots, A_n$ if:

1. *the probabilities on elements $A_1, \ldots, A_n$ are given by a single probability distribution μ, and $\Sigma_{D \in \{A_1,\ldots,A_n\}} \mu(D) = 1$,*

2. *for any two distinct elements A_j and A_l, they cannot be true at the same time, i.e., $A_j \wedge A_l = false$,*

3. *one element in the set has to be true given any circumstances, i.e., $\vee_{j=1}^{n} A_j = true$.*

[2] A similar definition is given in [Laskey and Lehner, 1989] in which this set is called an auxiliary hypothesis set.

The simplest probabilistic assumption set has two elements A and $\neg A$, denoted as $S_{A,\neg A}$.

For any two distinct and independent probabilistic assumption sets, denoted by $S_{A_1,\ldots,A_n}$ and $S_{B_1,\ldots,B_m}$, the unified probabilistic assumption set is defined as:

$$S_{A_1,\ldots,A_n,B_1,\ldots,B_m} = S_{A_1,\ldots,A_n} \otimes S_{B_1,\ldots,B_m} = \\ = \{(A_i, B_j) \mid A_i \in S_{A_1,\ldots,A_n}, B_j \in S_{B_1,\ldots,B_m}\},$$

where $\otimes$ means set product and $\mu((A_i, B_j)) = \mu_1(A_i) \times \mu_2(B_j)$. μ_1 and μ_2 are the probability distributions on $S_{A_1,\ldots,A_n}$ and $S_{B_1,\ldots,B_m}$, respectively.

Example 8.2

Assume that the five assumptions, Z, X, V, Y, and W in Example 8.1, are in different probabilistic assumption sets. If the environment for $node_6$ derived from justification $\{(node_2, node_3)\}$ is $\{\{X, V\}\}$, then the joint probabilistic assumption set for this environment is $S_{X,\neg X} \otimes S_{V,\neg V}$. Similarly, the joint probabilistic assumption set for environment $\{\{Y, W\}\}$ is $S_{Y,\neg Y} \otimes S_{W,\neg W}$.

$\diamondsuit$

Definition 8.2: *Full extension of a label*

Let E be a collection of different probabilistic assumption sets in a given ATMS. Let an environment of $node_n$ be $\{A, B, \ldots, C\}$ where $A, B, \ldots, C$ are in different probabilistic assumption sets $S_{A_1,\ldots,A_x}$, $S_{B_1,\ldots,B_y}$ and $S_{C_1,\ldots,C_z}$, respectively.

Set $\{A, B, \ldots, C\} \otimes S_{E_1,\ldots,E_t} \otimes \ldots \otimes S_{F_1,\ldots,F_j}$ (where $S_{E_1,\ldots,E_t}, \ldots, S_{F_1,\ldots,F_j} \in E \backslash (S_{A_1,\ldots,A_x} \cup S_{B_1,\ldots,B_y} \cup S_{C_1,\ldots,C_z})$) is called the full extension of the environment to all assumptions, or simply – the full extension of the environment.

When every environment in the label of $node_n$ has been fully extended to all assumptions, then the extended result is called the full extension of the label, denoted as $FL(n)$.

The principle of obtaining an extension of an environment is that if $A \wedge B \wedge \ldots \wedge C \rightarrow n$ is true, then $A \wedge B \wedge \ldots \wedge C \wedge (\vee_j E_j \mid E_j \in S_{E_1,\ldots E_t}) \rightarrow n$ should be true as well (where $S_{E_1,\ldots,E_t}$ is a probabilistic assumption set which is different from $S_{A_1,\ldots,A_x}, S_{B_1,\ldots,B_y}$ and $S_{C_1,\ldots,C_z}$).

To understand the idea behind this definition, we look at Example 8.1 again. There are 5 probabilistic assumption sets in that ATMS structure:

$S_{Z,\neg Z}$, $S_{X,\neg X}$, $S_{V,\neg V}$, $S_{Y,\neg Y}$ and $S_{W,\neg W}$. One environment of $node_6$ is $\{X, V\}$ which contains assumptions in two probabilistic assumption sets $S_{X,\neg X}$ and $S_{Y,\neg Y}$. Based on Definition 8.2, the full extension of this environment is

$$\{X, V\} \otimes S_{Z,\neg Z} \otimes S_{Y,\neg Y} \otimes S_{W,\neg W},$$

and the full extension of label $L(node_6)$ is

$$\{X, V\} \otimes S_{Z,\neg Z} \otimes S_{Y,\neg Y} \otimes S_{W,\neg W} \cup \{Y, W\} \otimes S_{X,\neg X} \otimes S_{V,\neg V} \otimes S_{Z,\neg Z}.$$

Similarly, we are able to calculate full extensions of labels for all other nodes.

Once we have obtained the full extension of the label of a node and have represented it in the disjunctive normal form, then each disjunctive component in the full extension contains the elements from all different probabilistic assumption sets and any two such disjunctive components are different. Such a full extension is the foundation for calculating beliefs of nodes when uncertainties are assigned to related assumptions.

Example 8.3

In Example 8.1, $node_6$ has label $\{\{X, V\}, \{Y, W\}\}$. Each of the two environments in this label involves two different probabilistic assumption sets. Although the probability of a node is defined as the probability of its label, the probability of $node_6$ cannot be obtained by calculating the probabilities of these two environments separately and then adding them together. Doing so may over-count the joint part in these two sets.

To calculate the belief in $node_6$, we will have to apply Definition 8.2 to each of these environments to get the full extensions, i.e.

$$S_{Z,\neg Z} \otimes \{X, V\} \otimes S_{Y,\neg Y} \otimes S_{W,\neg W},$$

and

$$S_{Z,\neg Z} \otimes S_{X,\neg X} \otimes S_{V,\neg V} \otimes \{Y, W\}.$$

The full extension of the label of $node_6$ is the union of these two sets:

$$(S_{Z,\neg Z} \otimes \{X, V\} \otimes S_{Y,\neg Y} \otimes S_{W,\neg W}) \cup (S_{Z,\neg Z} \otimes S_{X,\neg X} \otimes S_{V,\neg V} \otimes \{Y, W\}),$$

or

$$S_{Z,\neg Z} \otimes (\{X, V\} \otimes S_{Y,\neg Y} \otimes S_{W,\neg W} \cup S_{X,\neg X} \otimes S_{V,\neg V} \otimes \{Y, W\}).$$

If μ_X, μ_Y, μ_Z, μ_V, μ_W are the probability distribution on probabilistic assumption sets $S_{X,\neg X}$, $S_{Y,\neg Y}$, $S_{Z,\neg Z}$, $S_{V,\neg V}$, $S_{W,\neg W}$, then the degree of belief in this node is:

$$
\begin{aligned}
&Bel(node_6)\\
&= \mu_Z(S_{Z,\neg Z})(\mu_X(X)\mu_V(V)\mu_Y(S_{Y,\neg Y})\mu_W(S_{W,\neg W})\\
&\quad +\mu_X(S_{X,\neg X})\mu_V(S_{V,\neg V})\mu_Y(Y)\mu_W(W)\\
&\quad -\mu_X(X)\mu_V(V)\mu_Y(Y)\mu_W(W))\\
&= \mu_Z(S_{Z,\neg Z})(\mu_X(X)\mu_V(V) + \mu_Y(Y)\mu_W(W)\\
&\quad -\mu_X(X)\mu_V(V)\mu_Y(Y)\mu_W(W)).
\end{aligned}
$$

◇

In general, if the label of *nogood* is not empty, then the environments in this label should be deleted from the label of any other node, whenever these *nogood* environments appear in that label.

The label of a node is revised as

$$L(node) = L(node) \setminus L(nogood),$$

or

$$FL(node) = FL(node) \setminus FL(nogood).$$

The probability of a node is then changed as:

$$Bel(node) = \mu(FL(node) \setminus FL(nogood)).$$

For a premise node in an ATMS, such as N_9 in Table 8.1, it is not associated with any assumption(s). However, in this book, in order to create a unified procedure for extending labels of nodes, we require that each premise node is assigned to a distinct assumption and then is re-written into the form of assumed node. The probability of this distinct assumption is assigned to 1. For instance, if assumption D is assigned to node N_9 then node N_9 can be re-written as

$$N_9 :< q_1, \{\{D\}\}, \{(D)\} >$$

with $\mu(D) = 1$. The probabilistic assumption set is $S_{D,\neg D}$.

8.4 Conclusion

The main objectives of this chapter were to introduce the basic elements of an ATMS, to address the concepts of non-redundant justifications and environments, and to propose a method of calculating probabilities of statements (nodes).

Due to the fact that the label of a node consists of several environments in general, and these environments may share some common assumptions, the overlapped information in these environments has to be cancelled before the probability of a label can be correctly calculated.

Two new definitions, a *probabilistic assumption set* and a *full extension of a label*, are introduced in order to perform the calculation of probabilities. The concept of a *full extension of a label* enables an agent to transform a label into another form from which the calculation of probabilities can be easily done.

The discussion in this chapter sets the foundation for the study of relations between incidence calculus and the ATMS to be presented in the next chapter.

In [Liu, 1998], [Wells, *et al*, 1998a] and [Wells, *et al*, 1998b], the ATMS node structure has been adapted to represent network topologies. In the adapted ATMS structure, each physical device is represented by a node. The justifications of a node are used to identify those nodes which are vital in diagnosing whether the current node is faulty.

Chapter 9

Relations Between Extended Incidence Calculus and Assumption-Based Truth Maintenance System

In Chapter 7, we performed a comprehensive comparison between DS theory and generalized incidence calculus, under the perception that generalized incidence calculus is taken as a numerical approach for reasoning under uncertainty. However, as generalized incidence calculus can also be viewed as a pure symbolic reasoning mechanism, it would be interesting to see whether it has any formal links with other pure symbolic reasoning approaches, such as the ATMS.

Indeed, this chapter is devoted to the discussion of the formal relationship between generalized incidence calculus and the ATMS, a representative for symbolic reasoning approaches, based on the research results in [Liu and Bundy, 1996].

We first prove that managing labels of statements (nodes) in an ATMS is equivalent to producing incidence sets of these statements in generalized incidence calculus.

We then demonstrate that the set of justifications of a node is functionally equivalent to the *essential semantic implication set* of the same node in generalized incidence calculus.

As a consequence, generalized incidence calculus can provide justifications for an ATMS because essential semantic implication sets are discovered by the system automatically.

We also show that generalized incidence calculus provides a theoretical basis for constructing a probabilistic ATMS by associating proper probability distributions with assumptions. In this way, we can produce not only labels for all nodes, but also the probabilities of them in an ATMS. Moreover, different probability distributions in generalized incidence calculus are not necessarily to be independent. The 'nogood' environments can also be obtained automatically.

Therefore, generalized incidence calculus and the ATMS are equivalent in carrying out inferences at both the symbolic level and the numerical level, if the ATMS is integrated with a numerical method.

The research result to be presented in this chapter extends the result obtained in [Laskey and Lehner, 1989].

9.1 Review of Generalized Incidence Calculus

Incidence calculus was introduced in [Bundy, 1985], [Bundy, 1992] to deal with problems in pure numerical probabilistic reasoning. A special feature of this reasoning method is the indirect association of numerical uncertainty with formulae. In incidence calculus, probabilities are associated with the elements of a set of possible worlds and some formulae (called axioms) are associated with subsets of the set of possible worlds. The subset associated with each axiom, ϕ, contains those possible worlds which make the formula (axiom) true and this subset is normally called the incidence set of the formula. It is denoted as $i(\phi)$.

Given a GICT, $< \mathcal{W}, \mu, P, \mathcal{A}, i >$, for any formula $\phi \in \mathcal{L}(P)$, $i(\phi)$ is known if $\phi \in \mathcal{A}$; otherwise, the lower bound of its incidence set is calculated as

$$i_*(\phi) = \bigcup_{\psi \in \mathcal{A}, \psi \models \phi} i(\psi).$$

An agent's belief in a formula is regarded as the probability weight of the lower bound of its incidence set, i.e.,

$$Prob_*(\phi) = \mu(i_*(\phi)),$$

as defined in Chapter 3.

9.1.1 Essential Semantic Implication Sets in Incidence Calculus

Let us first give the following definitions:

Definition 9.1: *Semantic Implication Set*

Let ϕ be a formula in $\mathcal{L}(P)$ and $\mathcal{A}$ be a non-empty subset of $\mathcal{L}(P)$, if $\exists \psi \in \mathcal{A}$ such that $\psi \models \phi$ then ϕ is said to be semantically implied by ψ.

Let $SI_{\mathcal{A}}(\phi) = \{\psi \mid \psi \models \phi, \psi \in \mathcal{A}\}$, *then set* $SI_{\mathcal{A}}(\phi)$ *is called a semantic implication set of* ϕ, *given* $\mathcal{A}$.

For instance, if $\mathcal{A} = \{(a \to b) \wedge (b \to c), \ldots\}$, then $(a \to b) \wedge (b \to c)$ is in $SI_{\mathcal{A}}(a \to c)$ since $(a \to b) \wedge (b \to c) \models (a \to c)$.

Given a particular set, $\mathcal{A}$, the $SI_{\mathcal{A}}$ set of some formulae may be empty, such as, when $\mathcal{A} = \{a \to b, b \to c, a \to b \wedge b \to c\}$, then $SI_{\mathcal{A}}(e)$ is empty for $e \in \mathcal{L}(P)$.

Definition 9.2: *Essential Semantic Implication Set*

Let $ESI_{\mathcal{A}}(\phi)$ *be a subset of* $SI_{\mathcal{A}}(\phi)$. $ESI_{\mathcal{A}}(\phi)$ *is called an essential semantic implication set of* ϕ, *if it satisfies the condition*

$$\forall \psi \in ESI_{\mathcal{A}}(\phi) \text{ and } \forall \psi' \in SI_{\mathcal{A}}(\phi) \ \psi \not\models \psi' \text{ if } \psi \neq \psi'.$$

Given a $SI_{\mathcal{A}}(\phi)$ set, $ESI_{\mathcal{A}}(\phi)$ contains those 'biggest' formulae in $SI_{\mathcal{A}}(\phi)$. That is, for any formula $\psi \in ESI_{\mathcal{A}}(\phi)$, ψ does not imply any other formulae in $SI_{\mathcal{A}}(\phi)$.

We have the following property:

Proposition 9.1:

Given a GICT, $< \mathcal{W}, \mu, P, \mathcal{A}, i >$, *the essential semantic implication set of a formula is unique.*

Proof.

Obviously, the semantic implication set of any formula is unique, given $\mathcal{A}$.

Let $ESI_{\mathcal{A}}(\phi)$ and $ESI'_{\mathcal{A}}(\phi)$ be two different essential semantic implication sets of formula ϕ. Let ψ be a formula which is in $ESI_{\mathcal{A}}(\phi)$ but not in $ESI'_{\mathcal{A}}(\phi)$.

Because $\psi \in ESI_{\mathcal{A}}(\phi)$, for any formula $\psi' \in SI_{\mathcal{A}}(\phi)$, $\psi \not\models \psi'$ should hold.

On the other hand, as $\psi \notin ESI'_{\mathcal{A}}(\phi)$, there is at least one formula ψ'' ($\psi'' \in SI_{\mathcal{A}}(\phi)$) such that $\psi \models \psi''$.

According to Definition 9.2, $\psi \notin ESI_{\mathcal{A}}(\phi)$. So, there is a contradiction. Therefore, $ESI_{\mathcal{A}}(\phi) = ESI'_{\mathcal{A}}(\phi)$ and there is one and only one essential semantic implication set for any formula.

QED

It will be proved later that the essential semantic implication set of a formula is exactly the same as the set of justifications of that formula in an ATMS.

Example 9.1

Let $< \mathcal{W}, \mu, P, \mathcal{A}, i >$ be a GICT where $\mathcal{A}$ contains the following five inference rules and all the possible conjunctions of them:

$$r_1 : e \to d, \qquad r_2 : d \to b,$$

$$r_3 : b \to a, \qquad r_4 : d \to c,$$

$$r_5 : c \to a.$$

For any other formula, such as $e \to a$, the lower bound of its incidence set can be calculated using equation (3.3). $i_*(e \to a)$ is then calculated as:

$$i_*(e \to a) = \bigcup_{\phi \models (e \to a), \phi \in \mathcal{A}} i(\phi).$$

According to Definition 9.1, all the formulae ϕ in $\mathcal{A}$ satisfying the condition that $\phi \models (e \to a)$ are in the semantic implication set of $e \to a$.

Therefore, equation (3.3) can be restated as:

$$i_*(\psi) = \bigcup_{\phi \in SI_{\mathcal{A}}(\psi)} i(\phi).$$

In this example, there are in total seven axioms satisfying this requirement, so there are seven axioms in $SI_{\mathcal{A}}(e \to a)$. They are:

$$\begin{array}{l}
(e \to d) \land (d \to b) \land (b \to a), \\
(e \to d) \land (d \to c) \land (c \to a), \\
(e \to d) \land (d \to c) \land (c \to a) \land (d \to b), \\
(e \to d) \land (d \to c) \land (c \to a) \land (b \to a), \\
(e \to d) \land (d \to c) \land (c \to a) \land (d \to b) \land (b \to a), \\
(e \to d) \land (d \to b) \land (b \to a) \land (d \to c), \\
(e \to d) \land (d \to b) \land (b \to a) \land (c \to a).
\end{array}$$

However if we examine these seven axioms closely, we will find that only the first two axioms are necessary for calculating $i_*(e \to a)$. The rest are unnecessary as their incidence sets are included into the incidence sets of the first two axioms. Due to Definition 9.2, these two axioms are in the essential semantic implication set of $e \to a$ and they are the only elements in this set.

◇

Therefore, the following proposition is natural:

Proposition 9.2:

Let $SI_{\mathcal{A}}(\phi)$ and $ESI_{\mathcal{A}}(\phi)$ be the semantic implication set and the essential semantic implication set of ϕ, respectively, given a set of axioms $\mathcal{A}$.

Then, the following equation holds:

$$i_*(\phi) = i_*(SI_{\mathcal{A}}(\phi)) = i_*(ESI_{\mathcal{A}}(\phi)),$$

where $i_*(SI_{\mathcal{A}}(\phi)) = \bigcup_{\phi_j \in SI_{\mathcal{A}}(\phi)} i(\phi_j)$.

Proof.

Assume $\mathcal{A}$ is a set of axioms in a GICT. For formula ϕ, when $\phi \in \mathcal{A}$, we have:

$$\phi \in SI_{\mathcal{A}}(\phi), \phi \in ESI_{\mathcal{A}}(\phi), ESI_{\mathcal{A}}(\phi) = \{\phi\},$$

so that:

$$\begin{aligned} i_*(\phi) &= i(\phi), \\ i_*(SI_{\mathcal{A}}(\phi)) &= i(\phi) \cup (\bigcup_{\phi_j \in SI_{\mathcal{A}}(\phi)} i(\phi_j)) = i(\phi), \\ i_*(ESI_{\mathcal{A}}(\phi)) &= i(\phi) \cup (\bigcup_{\phi_j \in ESI_{\mathcal{A}}(\phi)} i(\phi_j)) = i(\phi), \end{aligned}$$

since $\bigcup_{\phi_j \in ESI_{\mathcal{A}}(\phi)} i(\phi_j) \subseteq i(\phi)$.

Therefore, the above equation holds.

When $\phi \notin \mathcal{A}$, then there is a list of formulae $\phi_1, \ldots, \phi_n \in \mathcal{A}$ (this set may be empty), each of which implies ϕ. So: $SI_{\mathcal{A}}(\phi) = \{\phi_1, \ldots, \phi_n\}$.

Let the elements in $ESI_{\mathcal{A}}(\phi)$ be $\psi_1, \ldots, \psi_m$, then, for ψ_j, there will be some formulae $\phi_{j'}$ (at least ψ_j itself) in $SI_{\mathcal{A}}(\phi)$ which make the following equation hold

$$\phi_{j'} \models \psi_j.$$

Let $SI_{\mathcal{A}}(\psi_j)$ be a set containing these $\phi_{j'}$, i.e. $SI_{\mathcal{A}}(\psi_j) = \{\phi_{j'} \mid \phi_{j'} \models \psi_j\}$, then we have $i_*(\psi_j) = i_*(SI_{\mathcal{A}}(\psi_j))$ as proved above.

Repeating this procedure for each formula in $ESI_{\mathcal{A}}(\phi)$, we obtain the following equation:

$$i_*(ESI_{\mathcal{A}}(\phi)) = \cup_{\psi_j \in ESI_{\mathcal{A}}(\phi)} i_*(SI_{\mathcal{A}}(\psi_j)).$$

To prove

$$i_*(SI_{\mathcal{A}}(\phi)) = i_*(ESI_{\mathcal{A}}(\phi)),$$

we need to prove

$$i_*(SI_{\mathcal{A}}(\phi)) = \cup_{\psi_j \in ESI_{\mathcal{A}}(\phi)} i_*(SI_{\mathcal{A}}(\psi_j)).$$

Assume that $i_*(SI_{\mathcal{A}}(\phi)) \setminus \cup_{\psi_j \in ESI_{\mathcal{A}}(\phi)} i_*(SI_{\mathcal{A}}(\psi_j)) = S \neq \emptyset$. Then, we have:

$S \neq \emptyset$ and $\exists w \in S$
$\Rightarrow w \in i_*(SI_{\mathcal{A}}(\phi)) \setminus \cup_{\psi_j \in ESI_{\mathcal{A}}(\phi)} i_*(SI_{\mathcal{A}}(\psi_j))$
$\Rightarrow (\exists \varphi)\varphi \in SI_{\mathcal{A}}(\phi), \varphi \notin ESI_{\mathcal{A}}(\phi), w \in i(\varphi)$
$\Rightarrow (\exists \varphi')\varphi' \in SI_{\mathcal{A}}(\phi), \varphi \models \varphi', \varphi' \notin ESI_{\mathcal{A}}(\phi)$
(otherwise $\varphi \in SI_{\mathcal{A}}(\varphi')$ and $\varphi \notin SI_{\mathcal{A}}(\phi)$)
$\Rightarrow (\exists \varphi'')\varphi'' \in SI_{\mathcal{A}}(\phi), \varphi' \models \varphi'', \varphi'' \notin ESI_{\mathcal{A}}(\phi)$
... (repeat this procedure until we find φ_t)
$\Rightarrow (\exists \varphi_t)\varphi_t \in SI_{\mathcal{A}}(\phi), \varphi_{t-1} \models \varphi_t, \varphi_t \notin ESI_{\mathcal{A}}(\phi)$ and $\nexists \varphi'_t, \varphi_t \models \varphi'_t$
... (since $\mathcal{A}$ is finite, this procedure will stop)
$\Rightarrow \varphi_t \notin ESI_{\mathcal{A}}(\phi)$ and $\varphi_t \in ESI_{\mathcal{A}}(\phi)$.

Conflict. So S is empty. Therefore, $i_*(SI_{\mathcal{A}}(\phi)) = i_*(ESI_{\mathcal{A}}(\phi))$ and $i_*(\phi) = i_*(SI_{\mathcal{A}}(\phi))$.

QED

Based on a GICT, the efficiency of calculating an incidence set for a formula is very much dependent on the efficiency of finding its semantic implication set as well as the essential semantic implication set.

9.1.2 Similarities of Reasoning Models in Generalized Incidence Calculus and the ATMS

In principle, if we view a set of possible worlds in generalized incidence calculus as a set of assumptions in an ATMS, and view the calculation of the incidence sets of formulae as the calculation of labels of nodes in the ATMS, then the two reasoning patterns are similar.

However, generalized incidence calculus has no such indications as justifications during its inference procedure. The implication relations among statements are discovered automatically.

The apparent similarity of these two reasoning patterns motivated us to further explore their formal relationship.

We will focus on the production of labels in the ATMS and the calculation of incidence sets in generalized incidence calculus.

We will prove that the two reasoning mechanisms are equivalent in producing dependent relations among statements.

Furthermore, as generalized incidence calculus can cope with statements with numerical degrees of belief, then generalized incidence calculus can be used as both a theoretical basis for the implementation of a probabilistic ATMS by providing both labels and degrees of belief of statements and an automatic reasoning model to provide justifications for an ATMS.

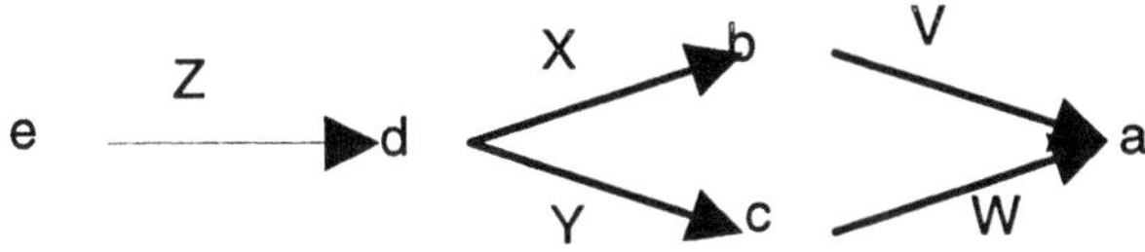

Figure 9.1: Semantic network of inference rules

9.2 Constructing Labels and Calculating Beliefs in Nodes Using Generalized Incidence Calculus

9.2.1 An Example

Before we proceed to the theoretical proof about the equivalence between these two reasoning mechanisms, we will employ an example, that has been mentioned in [Laskey and Lehner, 1989], to show how assumptions in the ATMS can be managed in the way that is done for a set of possible worlds in generalized incidence calculus.

This example will be represented using both an ATMS and generalized incidence calculus in order to illustrate the procedure of transforming an ATMS into the structure of generalized incidence calculus.

Example 9.2

Assume that five inference rules r_1, r_2, r_3, r_4, and r_5 are given initially as shown in Example 9.1, and that fact e is also observed. The semantics relationship of these five rules is shown in Figure 9.1.

The objective is to calculate the belief on other statements based on the given information, such as a.

Approach 1: Using an ATMS

Let the following nodes in an ATMS match the rules in Figure 9.1; these nodes are *assumed nodes*:

$$node_1 :< e \rightarrow d, \{\{Z\}\}, \{(Z)\} >,$$

$$node_2 :< d \rightarrow b, \{\{X\}\}, \{(X)\} >,$$

$$node_3 :< b \rightarrow a, \{\{V\}\}, \{(V)\} >,$$

$$node_4 :< d \rightarrow c, \{\{Y\}\}, \{(Y)\} >,$$

$$node_5 :< c \rightarrow a, \{\{W\}\}, \{(W)\} > .$$

Let $node_6$ stand for fact e; hence, it is a *premise node*:

$$node_6 :< e, \{\{\}\}, \{()\} > .$$

The assumed nodes are supported by assumptions. So, we create the following five *assumption nodes* for those five assumptions used above:

$$node_7 :< V, \{\{V\}\}, \{(V)\} >,$$

$$node_8 :< W, \{\{W\}\}, \{(W)\} >,$$

$$node_9 :< X, \{\{X\}\}, \{(X)\} >,$$

$$node_{10} :< Y, \{\{Y\}\}, \{(Y)\} >,$$

$$node_{11} :< Z, \{\{Z\}\}, \{(Z)\} > .$$

Finally, based on the assumed and promise nodes, a list of *derived nodes* is obtained as:

$$n_{12} :< d \rightarrow a, \{\{X, V\}, \{Y, W\}\}, \{(node_2, node_3), (node_4, node_5)\} >,$$

$$n_{13} :< e \rightarrow a, \{\{Z, X, V\}, \{Z, Y, W\}\}, \{(node_1, node_{12})\} >,$$

$$n_{14} :< a, \{\{Z, X, V\}, \{Z, Y, W\}\}, \{(node_{13}, node_6)\} >,$$

or

$$n_{13} :< e \rightarrow a, \{\{Z, X, V\}, \{Z, Y, W\}\}, \{(node_1, node_2, node_3),$$

$$(node_1, node_4, node_5\} >,$$

$$n_{14} :< a, \{\{Z, X, V\}, \{Z, Y, W\}\}, \{(node_1, node_2, node_3, node_6),$$

$$(node_1, node_4, node_5, node_6)\} >,$$

(after replacing $node_{12}$ and $node_{13}$ by their own justifications)

Obtaining labels of nodes is just the first step of calculating beliefs in nodes, as labels of nodes have to be further manipulated in order to perform the calculation (cf. [Pearl, 1988], [Laskey and Lehner, 1989]).

In the approach descriped in Chapter 8, the full extension of a label is essential. In order to do so, probabilistic assumption sets are required and some new assumptions need to be created whenever necessary.

For premise node e, a distinct assumption, say E, is required to be associated with it. $Node_6$ is then rewritten as $< e, \{\{E\}\}, \{(E)\} >$. Therefore, there are in total six probabilistic assumption sets. They are $S_{V,\neg V}$, $S_{W,\neg W}$, $S_{X,\neg X}$, $S_{Y,\neg Y}$, $S_{Z,\neg Z}$, $S_{E,\neg E}$.

The labels of derived nodes are obtained based on the justifications given by the problem solver, the premise nodes, and the assumed nodes. The label of proposition a is $L(a) = \{\{Z, X, V\}\{Z, Y, W\}\}$ and its full extension is:

$$FL(a) = S_{E,\neg E} \otimes \{Z\} \otimes (\{X, V\} \otimes S_{Y,\neg Y} \otimes \\ \otimes S_{W,\neg W} \cup S_{X,\neg X} \otimes S_{V,\neg V} \otimes \{Y, W\}).$$

If we assume that different probability distributions on different assumption sets are probabilistically independent, and these distributions give:

$$\mu_V(V) = 0.7, \quad \mu_W(W) = 0.8, \quad \mu_X(X) = 0.6.$$

$$\mu_V(\neg V) = 0.3, \quad \mu_W(\neg W) = 0.2, \quad \mu_X(\neg X) = 0.4.$$

$$\mu_Y(Y) = 0.75, \quad \mu_Z(Z) = 0.8, \quad \mu_E(E) = 0.0,$$

$$\mu_Y(\neg Y) = 0.25, \quad \mu_Z(\neg Z) = 0.2, \quad \mu_E(\neg E) = 0.0,$$

then the belief in node a is:

$$\begin{aligned} Bcl(a) &= \mu(FL(a)) \\ &= 1 \times 0.8 \times (0.6 \times 0.7 + 0.75 \times 0.8 - 0.6 \times 0.7 \times 0.75 \times 0.8) \\ &= 0.6144. \end{aligned}$$

A different calculation procedure, which produces the same result, can also be found in [Laskey and Lehner, 1989].

Approach 2: Using generalized incidence calculus

Now, let us see how this problem can be solved using generalized incidence calculus.

Suppose that we have the following six GICTs:

$$GICT_1, < S_{V,\neg V}, \mu_V, P, \{b \to a\}, \{i_1(b \to a) = \{V\}\} >,$$

$$GICT_2, < S_{W,\neg W}, \mu_W, P, \{c \to a\}, \{i_2(c \to a) = \{W\}\} >,$$

$$GICT_3, < S_{X,\neg X}, \mu_X, P, \{d \to b\}, \{i_3(d \to b) = \{X\}\} >,$$

$$GICT_4, < S_{Y,\neg Y}, \mu_Y, P, \{d \to c\}, \{i_4(d \to c) = \{Y\}\} >,$$

$$GICT_5, < S_{Z,\neg Z}, \mu_Z, P, \{e \to d\}, \{i_5(e \to d) = \{Z\}\} >,$$

$$GICT_6, < S_{E,\neg E}, \mu_E, P, \{e\}, \{i_6(e) = \{E\}\} >,$$

where $S_{V,\neg V}$, ..., $S_{Z,\neg Z}$ and $S_{E,\neg E}$ are probabilistic assumption sets, and μ_V, ..., μ_E are probability distributions on them, respectively.

We have ignored axioms *true* and *false* in all these GICTs in order to make statement simplier.

Since we assume that these sets $S_{X,\neg X}, \ldots, S_{Z,\neg Z}, S_{E,\neg E}$ are probabilistically independent, then the combination of the first five GICTs produces a new GICT:

$$GICT_7, < S_7, \mu_7, P, \mathcal{A}_7, i_7 >,$$

in which the joint set is $S_7 = S_{Z,\neg Z} \otimes S_{X,\neg X} \otimes S_{V,\neg V} \otimes S_{Y,\neg Y} \otimes S_{W,\neg W}$ with:

$$i_7(d \to b \wedge b \to a) = S_{Z,\neg Z}\{X\}\{V\}S_{Y,\neg Y}S_{W,\neg W},$$ [1]

$$i_7(d \to c \wedge c \to a) = S_{Z,\neg Z}\{Y\}\{W\}S_{X,\neg X}S_{V,\neg V},$$

$$i_7(d \to b \wedge b \to a \wedge d \to c \wedge c \to a) = S_{Z,\neg Z}\{X\}\{V\}\{Y\}\{W\},$$

$$i_7(e \to d \wedge d \to b \wedge b \to a) = \{Z\}\{X\}\{V\}S_{Y,\neg Y}S_{W,\neg W},$$

$$i_7(e \to d \wedge d \to c \wedge c \to a) = \{Z\}\{Y\}\{W\}S_{X,\neg X}S_{V,\neg V}.$$

If we let ϕ_1 and ϕ_2 stand for $(e \to d) \wedge (d \to b) \wedge (b \to a)$ and $(e \to d) \wedge (d \to c) \wedge (c \to a)$, respectively, then

$$i_7(\phi_1 \wedge \phi_2) = \{Z\}\{X\}\{V\}\{Y\}\{W\}.$$

A final GICT:

$$GICT_8, < S, \mu, P, \mathcal{A}, i >,$$

is obtained after combining $GICT_7$ with $GICT_6$[1], where:

$$i(e \wedge \phi_1) = S_{E,\neg E}\{Z\}\{X\}\{V\}S_{Y,\neg Y}S_{W,\neg W},$$

[1]The combination sequence does not affect the final result.

$$i(e \wedge \phi_2) = S_{E,\neg E}\{Z\}\{Y\}\{W\}S_{X,\neg X}S_{V,\neg V},$$

and

$$i(e \wedge \phi_1 \wedge \phi_2) = S_{E,\neg E}\{Z\}\{X\}\{V\}\{Y\}\{W\}.$$

Because $e \wedge \phi_1 \models a$, $e \wedge \phi_2 \models a$ and $e \wedge \phi_1 \wedge \phi_2 \models a$, then the following two equations hold:

$$\begin{aligned} i_*(a) &= i(e \wedge \phi_1) \cup i(e \wedge \phi_2) \cup i(e \wedge \phi_1 \wedge \phi_2) \\ &= S_{E,\neg E}\{Z\}\{X\}\{V\}S_{Y,\neg Y}S_{W,\neg W} \cup \\ &\quad \cup S_{E,\neg E}S_{X,\neg X}S_{V,\neg V}\{Z\}\{Y\}\{W\}, \end{aligned}$$

and

$$\begin{aligned} Prob_*(a) &= \mu(i_*(a)) \\ &= \mu(S_{E,\neg E}\{Z\}\{X\}\{V\}S_{Y,\neg Y}S_{W,\neg W} \cup \\ &\quad \cup S_{E,\neg E}S_{X,\neg X}S_{V,\neg V}\{Z\}\{Y\}\{W\}) \\ &= \mu(S_{E,\neg E}) \times \mu(\{Z\}\{X\}\{V\}S_{Y,\neg Y}S_{W,\neg W} \cup \\ &\quad \cup S_{X,\neg X}S_{V,\neg V}\{Z\}\{Y\}\{W\}) \\ &= \mu(S_{E,\neg E}) \times \mu(\{Z\}) \times \mu(\{X\}\{V\}S_{Y,\neg Y}S_{W,\neg W} \cup \\ &\quad \cup S_{X,\neg X}S_{V,\neg V}\{Y\}\{W\}) \\ &= \mu(S_{E,\neg E}) \times \mu(\{Z\}) \times (\mu(\{X\}\{V\}S_{Y,\neg Y}S_{W,\neg W}) + \\ &\quad + \mu(S_{X,\neg X}S_{V,\neg V}\{Y\}\{W\}) \\ &\quad - \mu(\{X\}\{V\}\{Y\}\{W\})) \\ &= 1 \times 0.8 \times (0.6 \times 0.7 \times 1 \times 1 + \\ &\quad + 1 \times 1 \times 0.75 \times 0.8 - 0.6 \times 0.7 \times 0.75 \times 0.8) \\ &= 0.6144. \end{aligned}$$

where μ is the final probability distribution in the final GICT.

So, our belief in a is also 0.6144.

Similarly, we can obtain $i_*(d \rightarrow a)$ and $i_*(e \rightarrow a)$ as:

$$\begin{aligned} i_*(d \rightarrow a) &= S_{E,\neg E}S_{Z,\neg Z}\{X\}\{V\}S_{Y,\neg Y}S_{W,\neg W} \cup \\ &\quad \cup S_{E,\neg E}S_{Z,\neg Z}\{Y\}\{W\}S_{X,\neg X}S_{V,\neg V}, \end{aligned}$$

and

$$\begin{aligned} i_*(e \rightarrow a) &= S_{E,\neg E}\{Z\}\{X\}\{V\}S_{Y,\neg Y}S_{W,\neg W} \cup \\ &\quad \cup S_{E,\neg E}\{Z\}\{Y\}\{W\}S_{X,\neg X}S_{V,\neg V}. \end{aligned}$$

If we compare the full extensions of nodes in the ATMS and the lower bounds of incidence sets on formulae, it is not difficult to find that the following equations hold:

$$\begin{aligned} i_*(d \rightarrow a) &= FL(d \rightarrow a), \\ i_*(e \rightarrow a) &= FL(e \rightarrow a), \\ i_*(a) &= FL(a). \end{aligned}$$

That is, the full extension of a node is exactly the same as the lower bound of incidence set of the corresponding formula, i.e., if $(a_1, a_2, \ldots, a_k) \in i_*(\phi)$, then $(a_1, a_2, \ldots, a_k) \in FL(\phi)$.

It is worth noting that the six GICTs initially specified are in fact produced from the assumed and premise nodes in the ATMS.

In the following we give a general procedure of encoding a list of ATMS nodes into equivalent GICTs.

9.2.2 Algorithm of Equivalent Transformation from an ATMS to Generalized Incidence Calculus

Algorithm E: Transformation from an ATMS to generalized incidence calculus

input: *an ATMS*
output: *a set of GICTs*

Step 1: $S_L \leftarrow \{all\ assumption\ nodes\}$

$S_{assumed} \leftarrow \{all\ assumed\ nodes\}$

$S_{premise} \leftarrow \{all\ premise\ nodes\}$

$S_{derived} \leftarrow \{all\ derived\ nodes\}$

$S_H \leftarrow S_{assumed} \cup S_{premise} \cup S_{derived}$

(S_L stands for the set of lower level nodes, and S_H for the set of high level nodes. Based on the higher level nodes, a set of atomic propositions P is established. A higher level node is either a proposition in P or a formula in $\mathcal{L}(P)$.)

Step 2: *from the set of assumption nodes, a list of probabilistic assumption sets* $S_{A_1,\ldots,A_m}, S_{B_1,\ldots,B_n}, \ldots,$ *can be constructed based on Definition 8.1.*

It is also assumed that these sets are probabilistically independent (if they are not independent, extended ATMS can not deal with them).

Step 3: *divide those assumed nodes into groups under the following conditions.*

for $n_j, n_l \in S_{assumed}$ **do**

  if $\exists A \in S_L$ **such that** $A \in E(n_j)$ *and* $A \in E(n_l)$

    where $E(n_j) \in L(n_j)$ *and* $E(n_l) \in L(n_l)$

  then n_j, n_l *are in the same group*

else if $\exists A \in S_L$, $\exists B \in S_L$ **such that** $A \in E(n_j)$ *and* $B \in E(n_l)$,

and A and B are in the same probabilistic assumption set,

then n_j, n_l *are in the same group*

if n_t *and* n_j *should be in the same group, and* n_j *and* n_l *should also be in the same group,*

then n_t, n_j *and* n_l *must be all in the same group.*

Step 4: $L_{GICT} \leftarrow \emptyset$

for each *group k* **do**

create $GICT_k, < \mathcal{W}_k, \mu_k, P, \mathcal{A}, i_k >$

$L_{GICT} \leftarrow L_{GICT} \cup \{GICT_k\}$

end *of do*

Function i_k *is defined as* $i_k(n_t) = L(n_t)$ *and* $i_k(n_t = n_l \wedge n_j) = L(n_l) \otimes L(n_j)$. *So* i_k, *defined on* $\mathcal{A}$, *is closed under* $\wedge$.

We further define $i_k(false) = \emptyset$ *and* $i_k(true) = \mathcal{W}_k$, *then* $< \mathcal{W}_k, p_k, P, \mathcal{A}, i_k >$ *is a GICT.*

The set of axioms $\mathcal{A}$ consists of assumed nodes in this group and all the possible conjunctions of them.

The set $\mathcal{W}_k$ is either a single probabilistic assumption set or the set product of several such sets if there is more than one probabilistic assumption set involved in the labels of these assumed nodes. For instance, if the label of node n_k is $\{\{A\},\{B\}\}$ and $S_{A,A_1,\ldots}, S_{B,B_1,\ldots}$ are different, then the set of possible worlds $\mathcal{W}_k$ should be $\mathcal{W}_k = S_{A,A_1,\ldots} \otimes S_{B,B_1,\ldots}$.

In the situation that the set of possible worlds is a joint space of several probabilistic assumption sets, labels of nodes need to be fully extended.

Following the above situation, if $S_{A,A_1,\ldots} = \{A, \neg A\}$ and $S_{B,B_1,\ldots} = \{B, \neg B\}$, the label of node n_t can be revised as:

$$\begin{aligned} L(n_t) &= \{\{A\} \otimes \{B, \neg B\}, \{A, \neg A\} \otimes \{B\}\} \\ &= \{\{\{A, B\}, \{A, \neg B\}\}, \{\{A, B\}, \{\neg A, B\}\}\} \\ &= \{\{A, B\}, \{A, \neg B\}, \{\neg A, B\}\}. \end{aligned}$$

Step 5: for each $n_j \in S_{premise}$ **do**

create $GICT_{n_j}$

$L_{GICT} \leftarrow L_{GICT} \cup \{GICT_{n_j}\}$

add *the set of possible worlds to the list of all probabilistic* *assumption sets.*

end *of do*

(For example, for premise e, a suitable GICT can be $< \{E, \neg E\}, \mu(V) = 1, P, \{e\}, i_j(e) = \{E\} >$. The added probabilistic assumption set must be different from any set in the list.)

Step 6: choose *two GICTs, G_j and G_l from L_{GICT}*

$G_{jl} \leftarrow G_j \oplus G_l$

$L_{GICT} \leftarrow \{G_{JL}\} \cup L_{GICT} \setminus \{G_j, G_l\}$

($\oplus$ means that two GICTs are combined.)

Step 7: if $\mid L_{GICT} \mid > 1$ **go to Step 6**

else for each *derived node* d_j, $i_*(d_j) = FL(d_j) \setminus FL(false)$.

($FL(d_j) \setminus FL(false)$ means deleting those conjunctive parts which appear in both $FL(d_j)$ and $FL(false)$.)

end *of algorithm*

Then, both the label set and the degree of belief in a node can be calculated in this combined GICT correctly.

Below, we will prove the correctness of the algorithm.

9.2.3 Formal Proof

In this section, we will prove that the above algorithm is sound.

Theorem 18 *Given an ATMS, there exists a set of GICTs such that the reasoning result of the ATMS is equivalent to the result obtained from the combination of these GICTs.*

For any node d_l in an ATMS, $FL(d_l) \setminus FL(false)$ is equivalent to the lower bound of the incidence set of formula d_l in the combined GICT, that is $FL(d_l) \setminus FL(false) = i_*(d_l)$.

The 'nogood' environments are equivalent to a subset of the set of possible worlds which causes conflicts, that is $FL(false) = \mathcal{W}_0$.

Proof.

Applying Algorithm E to a given ATMS, we get a list of GICTs. The nodes in the ATMS are divided into four sets, e.g., a set of assumption nodes, a set of assumed nodes, a set of premise nodes and a set of derived nodes.

Let us prove the three parts:

Part I: In order to carry out the proof, we need to reconstruct the justifications of derived nodes to ensure that justifications of derived nodes contain only assumed nodes or premise nodes. This can be done as follows. (We need to remember that justifications of a derived node contain all other possible nodes except assumption nodes. Each premise node has been attached to a distinct assumption.)

Given a derived node d_l, we choose a node d_j from its justifications. If node d_j is a derived node, then replace d_j with its justifications. For example if d_j is a derived node with justifications $\{(z_1, z_2)(z_3, z_4)\}$ and d_j appears in a justification of node d_l as $\{(\ldots, d_j, \ldots), \ldots\}$, then d_j is replaced with its justifications and the new justifications of d_l are $\{(\ldots, z_1, z_2, \ldots), (\ldots, z_3, z_4, \ldots), \ldots\}$.

We repeat this procedure until all nodes in the justifications of a derived node are either assumed nodes or premise nodes. As a consequence, an environment of a derived node contains only assumptions because labels of assumed and premise nodes contain only assumptions.

Part II: For any derived node d_l, suppose that its justifications are:

$$\{(a_1, a_2, \ldots), (b_1, b_2, \ldots), \ldots\},$$

then the conjunction of each justification of d_l implies d_l.

That is, $a_1 \wedge a_2 \wedge \ldots \models d_l$.

If we let $j(d_l) = \{a_1 \wedge a_2 \wedge \ldots, b_1 \wedge b_2 \wedge \ldots, \ldots\}$, then $j(d_l) \models d_l$. The label of d_l will be:

$$(L(a_1) \otimes L(a_2) \otimes \ldots) \cup (L(b_1) \otimes L(b_2) \otimes \ldots) \cup \ldots$$

Part III: After generating a language set from higher level nodes, a list of GICTs (assume m theories in total) can be constructed from assumed and premise nodes based on **steps 4 and 5** in Algorithm E. Any two sets of possible worlds of such theories are required to be probabilistically independent and all of them can be combined using Proposition 5.1 from Chapter 5, and the subset of possible worlds which leads to contradictions is W_0.

Now we examine the effect of combining these GICTs.

First of all, let $(n_1, n_2, \ldots, n_j)$ be a justification of a derived node, d_l, (we have ensured that these nodes are either assumed nodes or premise nodes). Nodes $(n_1, \ldots, n_l)$, re-written as $(n_{11}, \ldots, n_{1m_1}, \ldots n_{t1}, \ldots, n_{tm_t})$, are divided into t groups based on **Step 3**. Each group contains either a set of assumed nodes or a single premise node, and each group matches to a unique GICT.

Secondly, combining these t GICTs we obtain:

$$GICT_t, < \mathcal{W}_1, \mu'_1, P, \mathcal{A}'_1, i'_1 >, \tag{9.1}$$

with:

$$i'_1(n_1 \wedge n_2 \wedge \ldots \wedge n_j) = i_1(n_{11} \wedge \ldots \wedge n_{1m_1}) \otimes \ldots$$

$$
\begin{aligned}
&\ldots \otimes i_t(n_{t1} \wedge \ldots \wedge n_{tm_t}) \setminus W_1' = \\
&= (L(n_{11}) \otimes \ldots \otimes L(n_{1m_1}) \otimes \ldots \\
&\ldots \otimes (L(n_{t1}) \otimes \ldots \otimes L(n_{tm_t})) \setminus W_1' = \\
&= L(n_1) \otimes L(n_2) \otimes \ldots \otimes L(n_j) \setminus W_1',
\end{aligned}
$$

where W_1' is the subset of possible worlds which leads to contradictions after combing these t GICTs.

Thirdly, assume that $GICT_{m-t}$ is the GICT after combining the remaining $m-t$ GICTs:

$$GICT_{m-t}, < \mathcal{W}_2, \mu_2', P, \mathcal{A}_2', i_2' >, \tag{9.2}$$

where $\mathcal{A}_2' = \{y_1, y_2, \ldots, y_n\}$, and the subset of possible worlds leading to contradictions is $\mathcal{W}_2'$.

After combining $GICT_t$ and $GICT_{m-t}$ in (9.1) and (9.2), $\phi \wedge y_1, \phi \wedge y_2, \ldots, \phi \wedge y_n$ will be in the set of axioms of the new combined theory, i.e.

$$GICT, < \mathcal{W}_3, \mu_3', P, \mathcal{A}_3', i > . \tag{9.3}$$

Here, ϕ denotes $n_1 \wedge n_2 \wedge \ldots \wedge n_l$. Because $\phi \wedge y_j \models \phi$ and for any $\psi \wedge y_j \models \phi \wedge y_j$, $\psi \models \phi$, the following equation holds:

$$
\begin{aligned}
i_*(\phi) &= \textstyle\bigcup_j i(\phi \wedge y_j) \\
&= \textstyle\bigcup_j i_1'(\phi) \otimes i_2'(y_j) \setminus W_3' \\
&= i_1'(\phi) \otimes \textstyle\bigcup_j i_2'(y_j) \setminus W_3' \\
&= i_1'(\phi) \otimes (\mathcal{W}_2 \setminus W_2') \setminus W_3' \qquad \text{as } \textstyle\bigcup_j i_2'(y_j) = \mathcal{W}_2 \setminus W_2' \\
&= i_1'(\phi) \otimes \mathcal{W}_2 \setminus ((i_1'(\phi) \otimes W_2') \cup W_3') \\
&= (L(n_1) \otimes L(n_2) \otimes \ldots \otimes L(n_l) \setminus W_1') \otimes \mathcal{W}_2 \setminus ((i_1'(\phi) \otimes W_2') \cup W_3') \\
&= (L(n_1) \otimes L(n_2) \otimes \ldots \otimes L(n_l)) \otimes \mathcal{W}_2 \\
&\quad \setminus ((W_1' \otimes \mathcal{W}_2) \cup (i_1'(\phi) \otimes W_2') \cup W_3') \\
&= ((L(n_1) \otimes L(n_2) \otimes \ldots \otimes L(n_l)) \otimes \mathcal{W}_2) \setminus W_0,
\end{aligned}
$$

where W_3' is the set of possible worlds which leads to contradictions after combining the $GICT_t$ and $GICT_{m-t}$.

Therefore, the incidence function is i in the final GICT. W_0 is the total set of possible worlds causing conflict after combining all GICTs.

From the above analysis, we can see that if $(n_1, n_2, \ldots, n_j)$ is a justification of node d_l, then $\phi = n_1 \wedge \ldots \wedge n_l \models d_l$ holds in the ATMS. So, $n_1 \wedge \ldots \wedge n_l \models d_l$ also holds in generalized incidence calculus, and $i_*(\phi) \subseteq i_*(d_l)$.

In general, if node d_l has k justifications (e.g., (a_{11}, ..., a_{1x}), ..., (a_{k1}, ..., a_{kz})), then there will be k corresponding formulae ϕ_1, ϕ_2, ..., ϕ_k (e.g., $\phi_1 = a_{11} \wedge \ldots \wedge a_{1x}$) in generalized incidence calculus, where $\phi_j \models d_l$ and $i_*(\phi_j) \subseteq i_*(d_l)$ for $j = 1, \ldots, k$.

Therefore, $\bigcup_{\phi_j} i_*(\phi_j) \subseteq i_*(d_l)$.

The environments obtained from these k justifications are: $(L(a_{11})\otimes\ldots\otimes L(a_{1x})) \cup \ldots \cup (L(a_{k1} \otimes \ldots \otimes L(a_{ky})) \setminus L(false)$,

Step IV: Now we are ready to prove $FL(false) = W_0$.

In the ATMS, a *nogood* environment is derived if $false$ is proved. When c and $\neg c$ are both derived, then $L(c) \otimes L(\neg c)$ is a *nogood* environment. For any higher level node a, $(a, \neg a)$ is automatically recognized as a justification of node $false$ and $L(false) = nogood.$

Certainly, for an assumption A, $(A, \neg A)$ is also a justification of node $false$, but adding such justifications does not affect the result in our discussion, so in the following we only consider justifications of $false$ as $(a, \neg a)$.

Choosing a justification of node $false$, such as $(c, \neg c)$, $L(c)\otimes L(\neg c)$ should be in the label of *nogood*. When c or $\neg c$ is a derived node, we replace c or $\neg c$ with their labels.

Suppose that the justifications of c are $\{(z_1, z_2, \ldots), (x_1, x_2, \ldots), \ldots\}$ and the justifications of $\neg c$ are $\{(y_1, y_2, \ldots), \ldots\}$, then $\{(z_1, z_2, \ldots, y_1, y_2, \ldots), (x_1, x_2, \ldots, y_1, y_2, \ldots), \ldots\}$ will be justifications of $false$. Therefore:

$$(L(z_1)\otimes L(z_2)\otimes\ldots\otimes L(y_1)\otimes L(y_2)\otimes\ldots)\cup(L(x_1)\otimes L(x_2)\otimes\ldots\otimes L(y_1)\otimes L(y_2)\otimes\ldots)$$

are *nogood* environments.

Because $z_1\wedge z_2\wedge\ldots\wedge y_1\wedge\ldots = false$ and $x_1\wedge x_2\wedge\ldots\wedge y_1\wedge\ldots = false$, then we have $(L(z_1)\otimes L(z_2)\otimes\ldots\otimes L(y_1)\otimes\ldots)\cup(L(x_1)\otimes L(x_2)\otimes\ldots\otimes L(y_1)\otimes\ldots) \subseteq W_0$ based on **Step III** above.

Therefore, $FL(false) \subseteq W_0$.

The other way round, for any element $w \in W_0$ in the final GICT, there must be a formula $\phi_1 \wedge \phi_2 \wedge \ldots \wedge \phi_n = false$ and $w \in i(\phi_1 \wedge \phi_2 \ldots \wedge \phi_n)$. This formula is obtained from two axioms in the two GICTs being combined.

Therefore, there should exist a node z, the conjunction of some ϕ_j implies z and the conjunctions of remaining ϕ'_j implies $\neg z$, i.e., $z\wedge\neg z = \phi_1\wedge\ldots\wedge\phi_n = false$.

As a consequence, $L(\phi_1) \otimes \ldots \otimes L(\phi_n)$ are nogood environments and $w \in L(\phi_1) \otimes \ldots \otimes L(\phi_n)$. It is straightforward that w is in the full extension of $L(\psi_1) \otimes \ldots \otimes L(\psi_m)$. That is, $FL(false) \supseteq W_0$.

Finally, we have $FL(false) = W_0$.

Step V: Based on the result: $\bigcup_{\phi_j} i_*(\phi_j) \subseteq i_*(d_l)$ from the end of **Step III**, we have the following inequalities:

$$((L(a_{11}) \otimes \ldots \otimes L(a_{1x})) \otimes \ldots \otimes (L(a_{k1} \otimes \ldots \otimes L(a_{ky}))) \setminus W_0 \subseteq i_*(d_l),$$

and

$$FL(d_l) \setminus FL(false) \subseteq i_*(d_l).$$

On the other hand, for any $w \in i_*(d_l)$, there exist two axioms ψ_1 and ψ_2 (where $\psi_1 = \phi_1 \wedge \ldots \wedge \phi_n$ and $\psi_2 = \phi_{n+1} \wedge \ldots \wedge \phi_t$) from the two GICTs

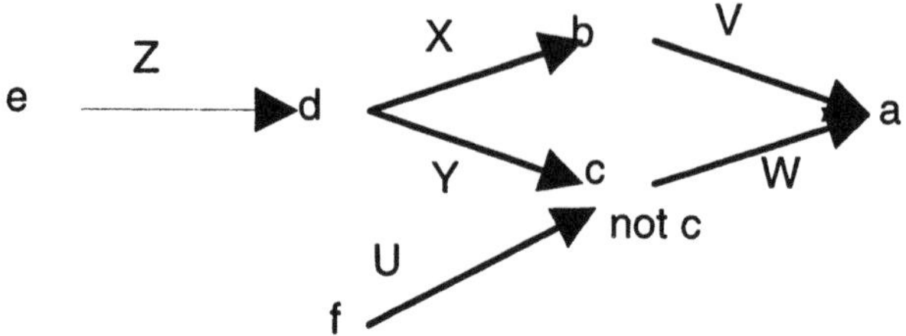

Figure 9.2: Extended semantic network of more inference rules

which satisfy conditions $\psi_1 \wedge \psi_2 \models d_l$ and $w \in i(\phi)$ (occasionaly., $\psi_1 \wedge \psi_2 = d_l$ if d_l is an axioms as well). $(\phi_1, \ldots \phi_t)$ then should be a justification of d_l, and $L(\phi_1) \otimes \ldots \otimes L(\phi_t) \setminus L(false)$ ahould be the environments of d_l.

Therefore w is in the full extension of $FL(d_l) \setminus FL(false)$. That is, $FL(d_l) \setminus FL(false) \supseteq i_*(d_l)$.

Finally, $FL(d_l) \setminus FL(false) = i_*(d_l)$.

QED

The side product of above proof is that $ESI(d_i) = \jmath(d_i)$. This equation can be proved using the definitions of $ESI(*)$, $SI(*)$ and $\jmath(*)$. $ESI(d_i) = \jmath(d_i)$ is a straightforward result of **Step III** and **Step V** as well as the definitions of $ESI(d_i)$, and justifications of a node.

Example 9.3

This example shows the method of dealing with conflicting information. Following the story in Example 9.2, suppose that we are told later that f is also observed and there is a rule $f \rightarrow \neg c$ with degree 0.8 in the knowledge base.

Then three more nodes in the ATMS will be added:

- n_{15}: *assumed node:* $< f \rightarrow \neg c, \{\{U\}\}, \{(U)\} >$,
- n_{16}: *premise node:* $< f, \{\{\}\}, \{()\} >$,
- n_{17}: *assumption node:* $< U, \{\{U\}\}, \{(U)\} >$, and
- *probability assumption sets:* $S_{U,\neg U}$, $S_{F,\neg F}$.

Here, $S_{F,\neg F}$ is created for premise node f, with $\mu_F(F) = 1$. For set $S_{U,\neg U}$, $\mu_U(U) = 0.8$ and $\mu_U(\neg U) = 0.2$. The node for f is changed as:

$$n_{16} :< f, \{\{F\}\}, \{(F)\} > .$$

In this ATMS, one environment of node c is $\{E, Z, Y\}$ and one environment of node $\neg c$ is $\{F, U\}$. So the *nogood* environment is $\{E, X, Y, F, U\}$.

The belief in node a needs to be re-calculated in order to re-distribute the weight of conflict to other nodes. The new belief in node a is 0.366 once its label has been fully extended.

In generalized incidence calculus, two more GICTs will be constructed from the assumed node $f \to \neg c$ and the premise node f.

Combining these two GICTs with the final one we obtained in Example 9.2, we have $W_0 = \{UZY\}^2$, $i_*(a) = \{ZXV \cup ZYW\} \setminus W_0$. Therefore, $\mu(\{UZY\}) = 0.48$ which is the weight of conflict and $Prob_*(a) = \mu(\{ZXV \cup ZYW\}) \setminus \{UZY\}) = 0.366$ which is our belief in a.

$\diamond$

Using both methods we obtained the results that are consistent with that given in [Laskey and Lehner, 1989].

9.2.4 Comparison with Laskey and Lehner's Work

The work described in this chapter has some similarity to that of Laskey and Lehner [Laskey and Lehner, 1989] in which the authors draw an equivalent relation between DS theory and an ATMS.

The key ideas in [Laskey and Lehner, 1989] are to create auxiliary elements between a set of statements and their numerical assignments, and to associate numerical assignments with auxiliary elements. The auxiliary elements are exactly the same as the set of possible worlds in generalized incidence calculus and the set of assumptions in an ATMS.

Both our and Laskey and Lehner's [Laskey and Lehner, 1989] work tried to group assumptions into different sets and associate each set with a probability distribution. Both systems calculate labels and degrees of belief in nodes. Both systems have to normalize labels after a conflict is discovered and both of them obtain a total conflict weight.

However, a formal proof of the connections between generalized incidence calculus and the ATMS is provided here while Laskey and Lehner did not prove the claim about an equivalence between DS theory and the ATMS. As a consequence, the result obtained in this chapter may serve as a theoretical basis for some of the results presented in [Laskey and Lehner, 1989], for which the proofs were missing.

Comparison 1: In [Laskey and Lehner, 1989], after the label of a node is obtained, an algorithm is given to rewrite a label as a list of disjoint conjuncts

[2] In order to state the problem simply, we use UZY to stand for $\{U\}\{Z\}\{Y\}S_{X,\neg X}S_{W,\neg W}S_{V,\neg V}S_{E,\neg E}S_{F,\neg F}$.

of assumptions, in order to calculate the belief in this node. For instance, in Example 9.2 the label of node a is rewritten as $L(a) = \beta_1 \vee \beta_2$ where $\beta_1 = W \wedge Y \wedge Z$ and $\beta_2 = (V \wedge X \wedge Z \wedge \neg W) \vee (V \wedge X \wedge Z \wedge W \wedge \neg Y)$.

If we simplify the elements in the full extension of a label (i.e. using Z to replace $(Z \wedge \neg W) \vee (Z \wedge W)$), we can get exactly those β lists required in [Laskey and Lehner, 1989].

Comparison 2: In [Laskey and Lehner, 1989], when *nogood* environments are not empty, the beliefs in other nodes are calculated in the following way:

$$Bel(node) = \frac{\mu(label \cap \neg nogood)}{\mu(\neg nogood)} = \frac{\mu(label \cap \neg nogood)}{1 - \mu(nogood)}.$$

Given a particular node d_l, they suggest that all the *nogood* environments can be divided into two groups, $nogood_1$ and $nogood_2$, where $nogood_2$ has no overlap with either environments in $nogood_1$ or $L(d_l)$. Then, in the actual calculation *nogood* will be replaced by $nogood_1$.

They further claim that such replacement will not affect the final result, without a proof. We will prove below that this result is sound.

Theorem 19 *Given a node, d_l, if all the 'nogood' environments can be divided into two disjoint groups $nogood_1$ and $nogood_2$, such that $nogood_2$ has no overlapps with either $nogood_1$ or $L(d_l)$.*

Then, the following equation holds:

$$Bel(d_l) = \frac{\mu(L(d_l) \cap nogood)}{1 - \mu(nogood)} = \frac{\mu(L(d_l) \cap nogood_1)}{1 - \mu(nogood_1)}.$$

Proof.

If all the *nogood* environments can be divided into two disjoint groups in respect of environments in $L(d_l)$, then it is possible to divide all the corresponding GICTs into two groups based on **Step III** in Section 9.2.3.

The combination of GICTs in two groups produces two conflict sets, referred to as $nogood_1$ and $nogood_2$, respectively.

The final combination of these two GICTs will not produce any more conflict sets (if it does then the assumption that $nogood_1$ and $nogood_2$ are disjoint is wrong).

Assume that the two incidence functions are i_1 and i_2, respectively after combining two groups of GICTs. For formula ϕ, if the list of axioms making ϕ true are $x_1, x_2, \ldots, x_n$ in the first combined theory, then:

$$i_{1_{*(\phi)}} = \bigcup_j (i_1(x_j))$$

Assume that the list of all axioms for incidence function i_2 is $(y_1, y_2, \ldots, y_m)$, then combining i_1 and i_2 we have:

$$\begin{aligned} i_*(\phi) &= \cup_l(\cup_j i(x_l \wedge y_j)) \\ &= \cup_l(\cup_j i_1(x_l) \otimes i_2(y_j)) \\ &= \cup_l(i_1(x_l) \otimes \cup_j i_2(y_j)) \\ &= \cup_l(i_1(x_l) \otimes (\mathcal{W}_2 \setminus FL(nogood_2))) \\ &= (\cup_l i_1(x_l)) \otimes (\mathcal{W}_2 \setminus FL(nogood_2)) \\ &= i_{1_{*(\phi)}} \otimes (\mathcal{W}_2 \setminus FL(nogood_2)). \end{aligned}$$

So, $Prob_*(\phi) = \mu(i_*(\phi)) = \mu'_1(i_*(\phi)) \times \mu'_2(\mathcal{W}_2 \setminus FL(nogood_2)) = \mu'_1(i_{1_{*(\phi)}})$. Here μ'_1 and μ'_2 are probability distributions on $\mathcal{W}_1 \setminus nogood_1$ and on $\mathcal{W}_2 \setminus nogood_2$, respectively, after normalization.

That is, the lower bound of probability weight on ϕ is the same as that before the two GICTs were combined.

Therefore:

$$Bel(\phi) = \frac{\mu(L(\phi) \cap nogood_1)}{1 - \mu(nogood_1)}.$$

QED

Comparison 3: The major step in [Laskey and Lehner, 1989] is to create an auxiliary set for each belief function and let the auxiliary set carry the information provided by the belief function. So, the probability distribution on an auxiliary set which in turn gives the belief function on another set can be thought as the source for this belief function.

Therefore, the two auxiliary sets defined in this way should be DS-independent. Otherwise, these two belief functions cannot be combined by Dempster's rule and the result obtained in an ATMS cannot be compared with the result in DS theory.

However, in generalized incidence calculus, we do not need to make such an assumption. For dependent probabilistic assumption sets, as long as we can find their joint probabilistic assumption set, we can still combine them using the combination rule in Chapter 5.

If there are a number of probabilistic assumption sets and some of them are dependent, then we combine dependent probabilistic assumption sets first and then carry out the combination for the rest.

Example 9.4

This example demonstrates the main issue we discussed in **Comparison 2** shown above.

Assume that the ATMS network is extended as shown in Figure 9.3 by adding more nodes in it. When facts h and j are observed, both i and $\neg i$ will

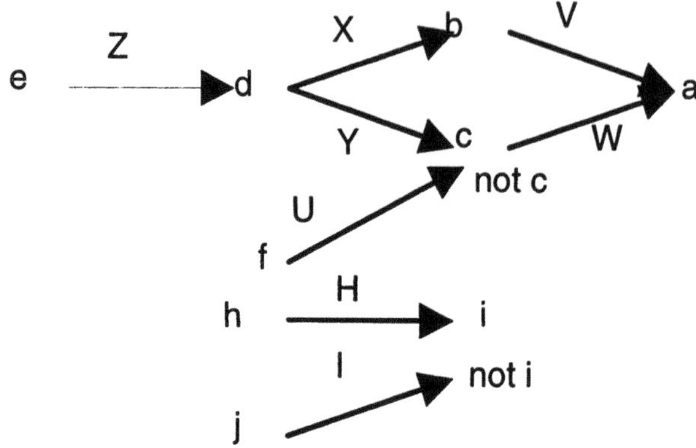

Figure 9.3: Extending the existing ATMS

be derived, then there will be a conflict. So, the total *nogood* environments are $\{UZY, HI\}$. If we let $nogood_1 = \{UZY\}$ and $nogood_2 = \{HI\}$, then $nogood_2$ has no overlap with $nogood_1$ and $L(a)$. Therefore, the belief in a should not be changed even when h and j are observed.

We have:

assumed nodes:	$< h \rightarrow i, \{\{H\}\}, \{(H)\} >$,
	$< j \rightarrow \neg i, \{\{I\}\}, \{(I)\} >$;
premise nodes:	$< h, \{\{\}\}, \{()\} >$,
	$< j, \{\{\}\}, \{()\} >$;
assumption node:	$< H, \{\{H\}\}, \{(H)\} >$,
	$< I, \{\{I\}\}, \{(I)\} >$;
probability	$S_{H,\neg H}$, $S_{I,\neg I}$,
assumption sets:	$S_{G,\neg G}$, $S_{L,\neg L}$.

where $S_{G,\neg G}$ and $S_{L,\neg L}$ are created for premises h and j, with $\mu_G(G) = 1$ and $\mu_L(L) = 1$.

The revised nodes for h and j are:

$$node_h :< h, \{\{G\}\}, \{(G)\} >,$$

and

$$node_j :< f, \{\{L\}\}, \{(L)\} > .$$

When considering this problem in generalized incidence calculus, once we have encoded the new assumed and premise nodes into GICTs, the combination of these GICTs produces a new GICT with conflict set $W_0' = \{HI\}$.

The further combination of this new GICT with the final GICT in Example 9.3 gives the total impact of all evidence. In this final GICT, it is possible to calculate beliefs on any nodes. For instance, $Prob'_*(a) = Prob_*(a) = 0.366$, with the total weight of conflict being:

$$\begin{aligned}
&\mu(FL(UZY \cup HI)) = \\
&= \mu_U(U)\mu_Z(Z)\mu_Y(Y) + \mu_H(H)\mu_I(I) + \\
&\quad -\mu_U(U)\mu_Z(Z)\mu_Y(Y)\mu_H(H)\mu_I(I) \\
&= 0.48 + \mu_H(H)\mu_I(I) - 0.48\mu_H(H)\mu_I(I).
\end{aligned}$$

$\diamond$

Therefore, in generalized incidence calculus, the correct result can still be achieved, without dividing *nogood* environments into different groups.

9.3 Generalized Incidence Calculus Can Provide Justifications for the ATMS

In the previous sections we discussed the formal relations between generalized incidence calculus and the ATMS. The major similarity of the two reasoning mechanisms is that the justifications in an ATMS are equivalent to the essential semantic implication sets in generalized incidence calculus. As a result, the labels of nodes are equivalent to the incidence sets of the corresponding nodes.

However, a difference between these two reasoning patterns is that the justifications are assigned by the designers in an ATMS while essential semantic implication sets are discovered automatically in generalized incidence calculus.

Therefore, the whole reasoning procedure in generalized incidence calculus is automatic while the one in an ATMS is semi-automatic.

The procedure of discovering semantic implication sets in generalized incidence calculus can be regarded as a tool to provide justifications for an ATMS.

The application of this procedure to an ATMS can release a system designer from the task of assigning justifications and this procedure can guarantee the derived justifications are non-redundant.

A problem arising from this procedure is that it is slow to find all essential semantic implication sets. If it is possible to have a fast algorithm for this procedure, then an ATMS can be established and generalized automatically without a designer's involvement.

We examine Example 9.2 in [Laskey and Lehner, 1989] in a different way here. Assume that our objective in Example 9.2 is to calculate the impact on a when e is observed. Because there is no direct effect from e on a, then a diagram shown as Figure 9.1 is created to build a link between e and a.

In order to infer a, the justifications for node $e \rightarrow a$ are essential to be given in an ATMS. Assume that the information carried by this diagram is denoted as $I_{initial}$ and the information specifying justifications is denoted as $I_{justifications}$, then in an ATMS we have:

$$I_{initial} \cup I_{justifications} \Rightarrow L(e \rightarrow a). \quad (9.4)$$

Here, notation $A \Rightarrow B$ means that from the information carried by A, it is possible to infer the information carried by B through some logical methods. $I_{justifications}$ may either contain the justifications for node $e \rightarrow a$ only or consists of more justifications for the assisting nodes (such as $e \rightarrow b$). $I_{justifications}$ is the extra information required in an ATMS in order to carry out system inferences.

Given the same initial information carried by $I_{initial}$, generalized incidence calculus does inferences without requiring any more information. The inference procedure produces:

$$I_{initial} \Rightarrow i_*(e \rightarrow a) \cup ESI(e \rightarrow a).$$

From the information in $I_{initial}$ it is possible to obtain both the lower bound of the incidence set and the inference paths for a node. The essential semantic implication set for a node contains exactly the justifications for the same node. Therefore, the extra information required by an ATMS can be supplied by generalized incidence calculus as an output.

Equation (9.4) will then be changed to:

$$I_{initial} \cup ESI(e \rightarrow a) \Rightarrow L(e \rightarrow a),$$

which takes the output from generalized incidence calculus as an input in the ATMS.

So, we can abstract out essential semantic implication sets for all necessary formulae and assign them to the corresponding nodes as justifications. In this way, a justification in an ATMS can be constructed automatically.

9.4 Conclusion

Integrating numerical approaches into an ATMS in order to deal with uncertainties has been a topic of research for some time.

In this chapter we paid particular attention to the possibility of constructing a probabilistic ATMS. Generalized incidence calculus is employed to study the underlining principles of a probabilistic ATMS and to examine the general structure for carrying out the corresponding inferences. The results show that it is possible to cope with uncertainties in an ATMS, when numerical values are attached to assumptions properly.

It is interesting to note a statement about the relationships between the ATMS and incidence calculus made by Pearl [Pearl, 1988]:

> "... in the original presentation of incidence calculus, propositions were not assigned numerical degrees of belief but instead were given a list of labels called *incidences*, representing a set of situations in which the propositions are true ... Thus, incidences are semantically equivalent to the ATMS notion of 'environments', and it is in this symbolic form that incidence calculus was first implemented by Bundy".

In the structure of generalized incidence calculus, both symbolic supporting relations among statements and numerical calculation of degrees of belief in different statements are explicitly described. For a specific problem, generalized incidence calculus can be either used as a symbolic based reasoning system or applied to deal with numerical uncertainties. This feature cannot be provided by pure symbolic or numerical approaches independently.

An advantage of using generalized incidence calculus to make inferences is that it does not require the problem solver to provide justifications. The whole reasoning procedure is performed automatically. The inference result can be used to produce the ATMS related justifications. The calculation of degrees of belief in nodes is based on the probability distributions on assumption sets which can either be dependent or independent.

Chapter 10

Conclusion

In this closing chapter we first present in some detail *rough sets theory* proposed by Pawlak (cf. [Pawlak, 1982]), and then discuss relationships between rough sets and incidence calculus.

In the second half of this chapter, some related work to incidence calculus, which was neither mentioned nor discussed in detail in previous chapters, will be presented. Following this, a short conclusion will be given to summarise the main contents in the book.

10.1 Rough Sets and Incidence Calculus

Rough sets theory was first proposed by Pawlak in 1982 [Pawlak, 1982]. Since then, the theory has been studied in many research papers and has become increasingly popular over the past 5 years, with successful applications in many areas, exemplified by machine learning, knowledge discovery, decision support systems, financial analysis, bioinformatics, etc. (e.g., [Slowinski, 1992], [Ziarko, 1993b], [Lin and Wildberger, 1994], [Polkowski and Skowron, 1998a], and [Polkowski and Skowron, 1998b], [Orlowska, 1998]).

An overview of rough sets can be found in [Düntsch and Gediga, 1999] and a valuable comprehensive tutorial is [Komorowski *et al.*, 1999].

Pawlak's book *Rough Sets* ([Pawlak, 1991]) gives a thorough study of theoretical aspects of rough sets.

In this section, we discuss set semantics of propositional logic in terms of rough sets. This discussion provides a basis for the investigation of relationships between rough sets and incidence calculus.

It should be noted that there have been some studies on the connections of formal logics with rough sets, as reported in [Lin and Wildberger, 1995], [Düntsch, 1997], [Wong, 1998], and several papers in [Orlowska, 1998]. Since we only intend to investigate relationships between rough sets and incidence calculus, then we will not review papers in that book. Readers interested in this topic are referred to those relevant publications.

10.1.1 Basics of Rough Sets

Rough sets theory is built upon set theory which is mainly about the coarsening and refinement of a *universe*, a non-empty, finite set of objects, based on *equivalence relations* at a different level of granularity.

Let U be a set, also called a universe, which is non-empty and contains a finite number of objects[1], and R be an equivalence relation on U. An equivalence relation is reflexive, symmetric and transitive.

An equivalence relation R on U divides the objects in U into a collection of disjoint sets with the elements in the same subset indiscernible. In other words, R induces a partition of U (or *classification*).

We denote each partition set, called an *equivalence class*, as W_l^R, and an element in W_l^R as w_{lj}^R. The family of all equivalence classes $\{W_1^R, \ldots, W_n^R\}$ is denoted as U/R. W_1^R and w_{lj}^R are simplified as W_l and w_{lj}, respectively, when there is no ambiguity about which equivalence relation R an agent is referring to.

Usually, we deal with several, rather than just one classification.

Let R and R' be two equivalence relations over U, $R \cap R'$ is a more refined equivalence relation, and $\cap$ can be understood as *and.*

The collection of equivalence classes of $R \cap R'$ is

$$U/(R \cap R') = \{W_l^R \cap W_j^{R'} \mid W_l^R \in U/R, W_j^{R'} \in U/R', W_l^R \cap W_j^{R'} \neq \emptyset\}. \tag{10.1}$$

Given a collection of equivalence relations, $\mathcal{R} = \{R_1, \ldots, R_n\}$, on a universe U, $R_1 \cap R_1 \cap \ldots \cap R_n$ is usually denoted as $IND(\mathcal{R})$ [Pawlak, 1991].

We have the following definition:

Definition 10.1: *Information system*

A structure $(U, \Omega, V_a)_{a \in \Omega}$ *is called an information system where:*

1. *U is a finite set of objects,*

2. *Ω is a finite set of primitive attributes describing objects in U,*

3. *For each $a \in \Omega$, V_a is the collection of all possible values of a.*

   Attribute a also defines a function:

$$a : U \to V_a$$

such that $\forall u \in U$, $\exists x \in V_a$, *that* $a(u) = x$.

[1] This assumption will not diminish the generality of properties of rough sets.

For a subset $Q \in \Omega$, function R_Q defined by:

$$u_1 R_Q u_2 \Leftrightarrow \forall a \in Q, a(u_1) = a(u_2)$$

is an equivalence relation.

When Q consists of only one attribute a, i.e., $Q = \{a\}$, then R_Q is called an *elementary equivalence relation.* In [Pawlak, 1991], each equivalence class in an elementary equivalence relation is referred to as an *elementary concept.* Any other non-elementary equivalence relation R_Q can be represented by a set of elementary equivalence relations using the following expression:

$$R_Q = \cap_{a \in Q} R_{\{a\}}.$$

Example 10.1

Assume that U is a universe containing 12 members of staff in a particular department (see Table 10.1). Let R be a relation on U which divides U into two disjoint sets, one with those members of staff who are also course directors and the other with the remaining members of staff.

Then, R is an equivalence relation. Similarly, relation R', which divides U into disjoint sets $\{u \in U$, Sex(u)=Male$\}$, $\{u \in U$, Sex(u)=Female$\}$, is also an equivalence relation.

The equivalence classes generated by R and R' are:

$$U/R = \{\{L1, L3, L5, L6, L10\}, \{L2, L4, L7, L8, L9, L11, L12\}\},$$

and

$$U/R' = \{\{L2, L5, L6, L7, L8, L10, L11, L12\}, \{L1, L3, L4, L9\}\}.$$

Through $R \cap R'$, it is possible for an agent to answer a query like 'list all female course directors'.

$R = R_{\{Cdirector\}}$ and $R' = R_{\{Sex\}}$ are elementary equivalence relations, but $R \cap R'$ is not.

◊

Next, we have:

Definition 10.2: *R-definable set*

Let U be a universe and R be an equivalence relation on U.

For a subset X of U, if X is the union of some W_l^R, then X is called R-definable; otherwise X is R-undefinable.

And we have the following property:

Table 10.1: A sample data table

U	Sex	Cdirector	Room#
L1	Female	Yes	EH02
L2	Male	No	EH03
L3	Female	Yes	EH04
L4	Female	No	ED02
L5	Male	Yes	ED05
L6	Male	Yes	EF02
L7	Male	No	EF03
L8	Male	No	EF04
L9	Female	No	EH06
L10	Male	Yes	EH07
L11	Male	No	EH01
L12	Male	No	EH09

Proposition 10.1:

Let R_1 and R_2 be two equivalence relations. Then, every equivalence class in U/R_1 or U/R_2 is $R_1 \cap R_2$-definable.

This proposition can be easily proved based on equation (10.1). In fact, for every $W_l^{R_1} \in U/R_1$ (or similarly $W_j^{R_2} \in U/R_2$),

$$W_l^{R_1} = \cup\{W_l^{R_1} \cap W_j^{R_2} \mid \forall W_j^{R_2} \in U/R_2\}$$

which is $R_1 \cap R_2$-definable.

More generally, we have the following two propositions:

Proposition 10.2:

Let $\mathcal{R} = \{R_1, \ldots, R_n\}$ be a collection of n equivalence relations.

Then, the subsets:

$$\bigcap_{1 \le l \le n} W_j^{R_l}$$

and

$$\bigcup_{1 \le l \le n} W_j^{R_l}$$

where $W_j^{R_l} \in U/R_l$ and $1 \le l \le n$ are all $IND(\mathcal{R})$-definable.

Proposition 10.3:

Let $\mathcal{R} = \{R_1, \ldots, R_n\}$ be a collection of n equivalence relations.

Then, any subset $X \subseteq U$ obtained by applying $\cap$ and $\cup$ to some equivalence classes in any $U/\mathbf{R}$ (where $\mathbf{R} \subseteq \mathcal{R}$) is $IND(\mathcal{R})$-definable.

Both of these two propositions can be proved based on Proposition 10.1 and equation (10.1). We leave the proofs of these propositions to the reader.

Any subset $X \subseteq U$ is called a *concept* [Pawlak, 1991]. Proposition 10.3 gives all the concepts that are definable under equivalence relation $IND(\mathcal{R})$.

For any subset X of U, we can also use two subsets of U to describe it as follows:

$$\underline{R}X = \cup\{W_l^R \mid W_l^R \subseteq X\},$$

and

$$\overline{R}X = \cup\{W_l^R \mid W_l^R \cap X \neq \emptyset\}.$$

When X is R-definable, then $\underline{R}X = \overline{R}X = X$.

Subsets $\underline{R}X$ and $\overline{R}X$ are called an *R-lower* and an *R-upper approximations* of X, respectively.

$\underline{R}X$ contains those objects which definitely prove the concept represented by X. $\overline{R}X$ contains those objects which possibly prove the concepts represented by X.

An information system (or a set of knowledge bases if every pair (U, R) is called a knowledge base) is represented as a data table in rough sets, like Table 10.1. Each data table has a number of rows labelled by objects (or states, processes, etc.) and columns by attributes. Each attribute defines an elementary equivalence relation and each equivalence class of the relation is uniquely identified by an attribute value.

If we use IF–THEN rules to state the relations among the attributes in a data table, some attributes can be described as *conditions* and some called *decisions*. A data table is called a *decision table* if condition and decision attributes are clearly distinguished.

When a new column 'organize course meetings' (just 'meeting' in short) is added to the above data table to indicate who is responsible for organizing course committee meetings, the relationship between 'course director' and 'organize course meetings' may then be established. Table 10.1 is therefore extended to a decision table as shown in Table 10.2.

Data and decision tables are the basic means of carrying information in rough sets, on which various manipulations are to be performed to derive useful results or knowledge.

10.1.2 Set Semantics of Propositional Logic

In general, equivalence relations on a universe define different partitions. For a specific partition, each equivalence class in the partition spells out a concept which can be characterized by a proper proposition.

Table 10.2: A decision table

U	Sex	Cdirector	Room#	Meeting
L1	Female	Yes	EH02	Yes
L2	Male	No	EH03	No
L3	Female	Yes	EH04	Yes
L4	Female	No	ED02	No
L5	Male	Yes	ED05	Yes
L6	Male	Yes	EF02	Yes
L7	Male	No	EF03	No
L8	Male	No	EF04	No
L9	Female	No	EH06	No
L10	Male	Yes	EH07	Yes
L11	Male	No	EH01	No
L12	Male	No	EH09	No

For instance, let R and U be the same as in Example 10.1 and let $P = \{q_1, q_2\}$ be a set of atomic propositions, where q_1 and q_2 stand for 'A lecturer is a course director' and 'A lecturer is not a course director' respectively. Then, the objects in equivalence class W_1 in U/R make q_1 true and the remaining objects in equivalence class W_2 prove q_2 true.

For objects in U, detailed in a data table, there is a set of primitive attributes, $\Omega = \{a_1, \ldots, a_n\}$ describing them. A particular object u in fact corresponds to a vector $< a_1 = x_1, \ldots, a_n = z_n >$ in the corresponding data table, where x_1 is the value that attribute a_1 is given, and z_n is the value assigned to a_n within this tuple. Let V_{al} be the domain of attribute a_l, a collection of all possible values that attribute a_l can take, then it is required that V_{al} has not omitted any potential values and a_l can take only one value at any particular time. In other words, V_{al} is a frame of discernment for a question in DS theory.

We have the following definition:

Definition 10.3: Valuation Function *val*

Let U be a non-empty universe with a finite number of objects, P be a finite set of atomic propositions. Function val given as:

$$val : U \times P \rightarrow \{true, false\} \tag{10.2}$$

is called a valuation function, and it assigns either true or false to every ordered pair (u, q) where $u \in U$ and $q \in P$.

$val(q, u) = true$, denoted as $(M, u) \models q$, can be understood as *q is true at (M, u)*, where $M = (U, R, val)$ is a Kripke structure ([Wong, 1998],

[Fagin *et al.*, 1995]), and R is a binary relation (e.g., an equivalence relation) on U.

Based on *val*, another mapping function $v : P \to 2^U$ can be derived as:

$$v(q) = \{u \mid u \in U, (M, u) \models q\}, \tag{10.3}$$

where $u \in v(q)$ is interpreted as q holds at state u.

Function v can be extended to a mapping $v : \mathcal{L}(P) \to 2^U$ as follows. For any formulae ϕ and ψ in $\mathcal{L}(P)$, there hold:

$$v(\phi \wedge \psi) = \{u \mid u \in U, ((M, u) \models \phi) \wedge ((M, u) \models \psi)\}, \tag{10.4}$$

$$v(\phi \vee \psi) = \{u \mid u \in U, ((M, u) \models \phi) \vee ((M, u) \models \psi)\}, \tag{10.5}$$

$$v(\neg \phi) = \{u \mid u \in U, (M, u) \not\models \phi\}. \tag{10.6}$$

Therefore, the subset of U containing those objects supporting formula ϕ (non-atomic proposition) can be derived through the initial truth assignment *val*.

Although we have given the description of *propositional logic* in Section 2.1.1, and we have emphasised that an atomic proposition q is a proposition which cannot be further broken down into more simpler propositions, we will nevertheless use notations we have defined here to further clarify what we mean when we say q is an atomic proposition.

We have now the following definitions:

Definition 10.4: *Atomic Proposition*

A proposition q is called an atomic proposition, if there exits one and only one attribute $a \in \Omega$ in an information system $(U, \Omega, V_a)_{a \in \Omega}$, such that $\exists x \in V_a$, $v(q) = \{u \mid a(u) = x\}$.

For example, 'A female lecturer' and 'A course director' are atomic propositions. 'Female course directors' is not an atomic proposition, as there are two attributes (gender and course director) associated with this sentence.

Definition 10.5: *Equivalence Relations versus Subsets of Atomic Propositions*

Let $M = (U, R, val)$ be a Kripke structure where R is an elementary equivalence relation. Let P be a finite set of atomic propositions.

If there is a subset $P' = \{q_1, \ldots, q_n\}$ of P such that $U/R = \{v(q_1), v(q_2), \ldots, v(q_n)\}$ holds, then subset P' is said to be equivalent to R, denoted as $U/R = v(P')$.

$v(P')$ is defined as a collection of subsets of U, i.e., $v(P') = \{v(q_1), v(q_2), \ldots, v(q_n)\}$, for all $q_l \in P'$.

This definition suggests that there can be a subset of a set of atomic propositions P which is functionally equivalent to an elementary equivalence relation in terms of the partitioning of a universe, with respect to a particular aspect (attribute) of the objects in the universe.

Example 10.2

Let q_1 and q_2 be two atomic propositions, 'A female lecturer' and 'A male lecturer', respectively, and R' be an elementary equivalence relation defined in Example 10.1.

Given the objects in U as shown in Table 10.1, these two atomic propositions divide U into two disjoint subsets:

$$v(q_1) = \{u_1, u_3, u_4, u_9\},$$

and

$$v(q_2) = U \setminus v(p_1),$$

where $v(q_1) = W_1^{R'}$ and $v(q_2) = W_2^{R'}$.

Therefore, $v(P') = U/R$. $P' = \{q_1, q_2\}$ is equivalent to R in the context of classifying objects in Table 10.1 into two groups: male and female lectures.

◇

Now, we consider whether a formula in $\mathcal{L}(P)$ can be characterized by equivalence classes, in general.

Definition 10.6: *R-definable Formulae*

Let $M = (U, R, val)$ be a Kripke structure, where R is an elementary equivalence relation on U. Let P be a finite set of atomic propositions.

For a formula ϕ in $\mathcal{L}(P)$, if $v(\phi)$ defined in equation (10.3) is R-definable then ϕ is said to be an R-definable formula.

Otherwise, ϕ is an R-underfinable formula . Formulae true and false are always R-definable with $v(true) = U$ and $v(false) = \emptyset$.

We have now some relevant properties.

Theorem 20 *Given a finite set of atomic propositions P and a Kripke structure $M = (U, R, val)$, with R being an elementary equivalence relation on U, when $U/R = v(P)$ holds, then every formula in $\mathcal{L}(P)$ is an R-definable formula.*

Proof.

Let $U/R = \{W_1, W_2, \ldots, W_n\}$. As R is an elementary equivalence relation, there exits an attribute a with n possible values where each possible value x_l is related to a unique W_l.

Let $P = \{q_1, \ldots, q_n\}$, then $v(q_l) \cap v(q_j) = \emptyset$ when $l \neq j$, because $v(q_l) = W_l$ and $v(q_j) = W_j$ and $W_l \cap W_j = \emptyset$. This means that any two different atomic propositions q_l and q_j cannot both hold on the same objects, so that $q_l \wedge q_j = false$.

According to the construction of the basic element set of a set of propositions (cf. Definition 2.2), each basic element $\delta_l = q'_1 \wedge q'_2 \wedge \ldots \wedge q'_n$ out of P (where q'_l is either q_l or $\neg q_l$), is reduced to $\delta = q_l$ for a particular l because one and only one proposition, q_l, is true given any particular object (as $q_l \wedge \neg q_j = q_l$ when $l \neq j$).

Therefore, the basic element set $\mathcal{A}t$ of P is the same as P. As a result, any formula ϕ in $\mathcal{L}(P)$ is in the form $\vee_l q_l$ ($n \geq l \geq 1$).

Since

$$v(\phi) = v(\vee_l q_l) = \{u \mid u \in U, (M, u) \models (\vee_l q_l)\}$$

and $v(q_l) \cap v(q_j) = \emptyset$ when $l \neq j$, then

$$v(\phi) = \cup_l \{u \mid u \in U, (M, u) \models q_l\} = \cup_l v(q_l).$$

Subset $\cup_l v(q_l) = \cup_l W_l$ is R-definable, so is $v(\phi)$.

Therefore, ϕ is also R-definable.

QED

Theorem 21 *Let $R_1, \ldots, R_n$ be n elementary equivalence relations and P_1, $\ldots$, P_n be n subsets of a set of atomic propositions P with $U/R_l = v(P_l)$, for $l = 1, \ldots, n$.*

If $P' = \cup_l P_l$, then every formula in $\mathcal{L}(P')$ is $IND(\mathcal{R})$-definable.

Proof.

We assume that elements in P_1, P_2, $\ldots$, P_n are labelled as

$$\begin{aligned} P_1 &= \{q_{11}, q_{12}, \ldots, q_{1t1}\}, \\ p_2 &= \{q_{21}, q_{22}, \ldots, q_{2t2}\}, \\ &\ldots \\ P_n &= \{q_{n1}, q_{n2}, \ldots, q_{ntn}\}. \end{aligned}$$

For any formula $\phi \in \mathcal{L}(P')$, ϕ is first transformed into its disjunctive normal form.

Let $\phi = \vee(\wedge q'_{lj})$, where $q'_{lj} = q_{lj}$ or $q'_{lj} = \neg q_{lj}$, and $q_{lj} \in P_l$.

For each conjunctive component $\wedge q'_{lj}$, let $\wedge q'_{lj} = (\wedge q_{lj}) \wedge (\wedge \neg q_{ht})$. That is, each conjunctive component is divided into two disjoint sub-components where the first sub-component contains no logical connective $\neg$.

Then, any two distinctive formulae q_{lj} and $q_{l'j'}$ in the first sub-component, $(\wedge q_{lj})$, will come from two different subsets, P_l and P'_l. Otherwise, $q_{lj} \wedge q_{l'j'}$ is *false* (also see the proof of the above theorem), and this sub-component (as well as this conjunctive component) should not have been in the disjunctive normal form.

According to equations (10.4) – (10.6), the subset of objects supporting statement $\wedge q_{lj}$ is:

$$v(\wedge q_{lj}) = \{u \mid u \in v(q_{lj}),$$

for every q_{lj} in this sub-component.

Also, for each $v(q_{lj})$, there is a $W_j^{R_l}$ such that $v(q_{lj}) = W_j^{R_l}$, and therefore,

$$v(\wedge q_{lj}) = \{u \mid u \in W_j^{R_l},$$

for every q_{lj} in this sub-component.

Therefore, $v(\wedge q_{lj}) = \cap W_j^{R_l}$ which is $IND(\mathcal{R})$-definable (cf. Proposition 10.2).

Similarly, for sub-component $(\wedge \neg q_{ht})$, the subset of objects supporting it is $v(\wedge \neg q_{ht}) = \cap(U \setminus W_t^{R_h})$ (because for every $\neg q_{ht}$, there is a $W_t^{R_h}$ such that $v(\neg q_{ht}) = U \setminus W_t^{R_h}$) which is also $IND(\mathcal{R})$-definable.

Therefore, $v((\wedge q_{lj}) \wedge (\wedge \neg q_{ht})) = \{u \mid u \in v(\wedge q_{lj}) \text{ and } u \in v(\wedge \neg q_{ht})\} = (\cap W_j^{R_l}) \bigcap (\cap(U \setminus W_t^{R_j}))$ is $IND(\mathcal{R})$-definable.

The above two steps together show that each conjunctive component of formula ϕ is $IND(\mathcal{R})$-definable.

Since $v(\phi) = \{u \mid u \in v(q_{lj}) \cap v(\neg q_{ht})$, for at least one conjunctive component $(\wedge q_{lj}) \wedge (\wedge \neg q_{ht})\}$, and $v(\phi)$ can be re-written as $\cup(v(q_{lj}) \cap v(\neg q_{ht}))$ $= \cup((\cap W_j^{R_l}) \cap (U \setminus W_j^{R_l}))$, which is also $IND(\mathcal{R})$-definable (cf. Proposition 10.3), then ϕ must be $IND(\mathcal{R})$-definable.

QED

Example 10.3

Let U be a set of objects containing a group of 10 students. Four attributes describing these students are given in detail in Table 10.3.

Let P_1 and P_2 be two subsets of a set of atomic propositions P as:

- $P_1 = \{q_{11}, q_{12}\}$={male students, female students}, and
- $P_2 = \{q_{21}, q_{22}, q_{23}\}$={students in Dr. White's group, students in Dr. Jones's group, students in Dr. Snow's group}.

Table 10.3: A data table containing 10 students

U	Sex	Major	Tutor group	Lab session
s1	Female	Bsc in CS	Dr. Jones	2
s2	Female	Bsc in CS	Dr. Snow	1
s3	Male	Bsc in CS	Dr. Jones	2
s4	Male	BENG	Dr. White	2
s5	Female	Bsc in CS	Dr. White	1
s6	Male	BENG	Dr. White	2
s7	Male	BENG	Dr. White	2
s8	Male	Bsc in CS	Dr. Jones	1
s9	Female	Bsc in CS	Dr. Jones	1
s10	Male	BENG	Dr. Snow	1

These two subsets of atomic propositions are equivalent to two elementary equivalence relations R_1 and R_2 with the equivalence classes:

$$U/R_1 = \{\{s3, s4, s6, s7, s8, s10\}, \{s1, s2, s5, s9\}\},$$

and

$$U/R_2 = \{\{s4, s5, s6, s7\}, \{s1, s3, s8, s9\}, \{s2, s10\}\}.$$

The following formulae:

ϕ =female students in Dr. White's group,

ψ =either female students or students in Dr. Jones's group,

φ =male students not in Dr. Snow's group,

which can be re-written into disjunctive normal forms:

$\phi = (q_{12} \wedge q_{21})$,

$\psi = (q_{12}) \vee (q_{22})$,

$\varphi = (q_{11} \wedge \neg q_{23})$

are all $R_1 \cap R_2$-definable.

The subsets of objects supporting these formulae – i.e., $v(\phi)$, $v(\psi)$, and $v(\varphi)$ – are, respectively:

$$\{s5\},$$

$$\{s1, s2, s3, s5, s8, s9\},$$

and

$$\{s3, s4, s6, s7, s8\}.$$

Now, we have:

Definition 10.7: *Complete Rough Logic Theory*

Structure (U, R, P, P', val) is called a complete rough logic theory, where:

- *$M = (U, R, val)$ is a Kripke structure, with R being an elementary equivalence relation on U.*
- *P' is subset of a finite set of atomic propositions P, with $U/R = v(P')$.*

Proposition 10.4:

Let (U, R, P, P', val) be a complete rough logic theory.

For any two formulae $\phi, \psi \in \mathcal{L}(P)$, the following equations hold:

$$\begin{aligned} v(\neg\phi) &= U \setminus v(\phi), \\ v(\phi \wedge \psi) &= v(\phi) \cap v(\psi), \\ v(\phi \vee \psi) &= v(\phi) \cup v(\psi), \\ v(\phi \rightarrow \psi) &= (U \setminus v(\phi)) \cup v(\psi). \end{aligned}$$

This proposition is easily provable based on the definition of function v and three equations from (10.4) to (10.6).

Each complete rough logic theory identifies an elementary equivalence relation R and a subset P' of a set of atomic propositions which are equivalence to R in the context of partitioning the objects in the given universe.

All the formulae in $\mathcal{L}(P')$ are $R-$definable. When a formula ϕ in $\mathcal{L}(P)$ is not R-definable, it is only possible to define the upper and lower approximations of $v(\phi)$:

$$\underline{v}(\phi) = \cup\{v(\psi) \mid \psi \models \phi, \psi \in \mathcal{L}(P')\}. \tag{10.7}$$

and

$$\overline{v}(\phi) = \cap\{v(\psi) \mid \phi \models \psi, \psi \in \mathcal{L}(P')\}, \tag{10.8}$$

respectively.

Valuation function val in a complete rough logic theory requires full information about every ordered pair (u, q) in the space $U \times P$, which subsequently determines a unique subset of U for every formula ϕ in $\mathcal{L}(P)$, i.e., $v(\phi)$.

When a universe U is very large, then it may not be practical to require function val being fully specified, but it may be quite reasonable to have some information about a particular elementary equivalence relation (R) and its corresponding equivalent subset of a set of atomic propositions (P').

In this case, $v(\phi)$ can be determined only when $\phi \in \mathcal{L}(P')$, as $v(\phi)$ can be represented using elements in U/R. That is, every formula in $\mathcal{L}(P')$ is R-definable and formulae in $\mathcal{L}(P) \setminus \mathcal{L}(P')$ is R-undefinable.

This leads to the following definition and proposition:

Definition 10.8: *Partial Rough Logic Theory*

Structure (U, R, P, P', val) such that:

- *U is a universe consisting of a finite number of objects,*
- *P is a finite set of atomic propositions,*
- *R is an elementary equivalence relation on U,*
- *Valuation function val is only partially specified on space $U \times P$,*
- *$P' \subset P$. For each $q_l \in P'$, $v(q_l) = W_l$ and W_l is in U/R,*

is called a partial rough logic theory.

According to this definition, Proposition 10.4 should be changed as follows:

Proposition 10.5:

Let (U, R, P, P', val) be a partial rough logic theory. For any two formulae $\phi, \psi \in \mathcal{L}(P')$, the following equations hold:

$$\begin{aligned} v(\neg\phi) &= U \setminus v(\phi), \\ v(\phi \wedge \psi) &= v(\phi) \cap v(\psi), \\ v(\phi \vee \psi) &= v(\phi) \cup v(\psi), \\ v(\phi \to \psi) &= (U \setminus v(\phi)) \cup v(\psi). \end{aligned}$$

Therefore, the set of objects supporting an atomic proposition in P', or more generally a formula ϕ in $\mathcal{L}(P')$, can be characterized by equivalence classes in U/R, given a partial rough logic theory.

In summary, a complete rough logic theory enables every $v(\phi)$ to be calculable, while a partial rough logic theory can only define some $v(\phi)$ precisely.

In practice, a partial rough logic theory is more natural than a complete rough logic theory because there are many reasons that *val* may not be available, such as U is too large, some attribute values of some objects are missing, details of data is not unavailability due to data security, etc.

10.1.3 Relationship Between Rough Sets and Incidence Calculus

The structure of a rough logic theory (complete or partial) proposed above is very similar to the definition of original incidence calculus theory in Chapter 2.

Given an original incidence calculus theory $< U, \mu, P, \mathcal{A}, i >$, the following equations hold for the original incidence function i on set $\mathcal{L}(\mathcal{A})$:

$$\begin{aligned} i(\neg\phi) &= U \setminus i(\phi), \\ i(\phi \wedge \psi) &= (\phi) \cap i(\psi), \\ i(\phi \vee \psi) &= i(\phi) \cup i(\psi), \\ i(\phi \rightarrow \psi) &= (U \setminus i(\phi)) \cup i(\psi). \end{aligned}$$

The properties of function i on language set $\mathcal{L}(\mathcal{A})$ is identical to the properties of function v on language set $\mathcal{L}(P)$ in Proposition 10.4, and on set $\mathcal{L}(P')$ in Proposition 10.5. $i(\phi)$ and $v(\phi)$ are two different methods of describing the same set of possible objects which make formula ϕ true. Function v is named as I in [Wong, 1998] in which I is directly referred to as an incidence function. However, partially defined function v in Proposition 10.5 is more close to the meaning of incidence function i, as i is also a partially defined function itself.

We have the following property:

Proposition 10.6:

Let (U, R, P, P', val) be a partial rough logic theory. Then, structure $< U, \mu,$ $P, \mathcal{L}(P'), v >$ is an original incidence calculus theory, where $\mu(u) = 1/ \mid U \mid$ for every $u \in U$. coup

Here, $\mid U \mid$ is the cardinality of set U, i.e., the total number of elements in U.

This proposition can be easily proved based on Definition 2.4 and Proposition 10.5.

Although it is always possible to define an original incidence calculus theory from a partial rough logic theory, it is not possible to do this the other way round. The main reasons are:

1. an incidence function is defined on an arbitrary subset of a language set, and

2. the initially assignment of incidences on axioms may not partition a universe.

In a partial rough logic theory, function v is initially defined on a subset of a set of atomic propositions, $P' = \{q_1, \ldots, q_n\}$, with the following equations:

$$\begin{aligned} v(q_l) &\neq \emptyset, \\ \cup_l v(q_l) &= U, \\ v(q_l) \cap v(q_j) &= \emptyset. \end{aligned}$$

These properties imply Proposition 10.7 to be shown below.

Proposition 10.7:

Let (U, R, P, P', val) be a partial rough logic theory. If, for every $q_l \in P'$, we define $ii(q_l) = v(q_l)$, then $ii : P' \cup \{false\} \to 2^U$ (with $ii(false) = \emptyset$) is a basic incidence assignment.

Proof.

Every ϕ_l in P' should be mapped to an equivalence class W_l^R in the way that W_l^R contains all the objects in U that make ϕ_l true. So ii is a mapping function $ii : P' \to 2^U$. ii also has the following properties:

$ii(q_l) \neq \emptyset$,

$ii(q_l) \cap ii(q_j) = \emptyset$, when $l \neq j$, and

$\cup_l ii(q_l) = U$.

We also define $ii(false) = \emptyset$, and $P' = P' \cup \{false\}$.
Therefore, ii is a basic incidence assignment (Definition 3.7).

QED

Proposition 10.8:

Let ii be a basic incidence assignment from P' to 2^U, where P' is a subset of a finite set of atomic propositions and U is a universe.

Then, $(U, R, P, P' \setminus \{false\}, val)$ is a partial rough logic theory with $W_l = ii(q_l)$ for every $q_l \in P' \setminus \{false\}$.

Proof.

If $ii(true) \neq \emptyset$, then an atomic proposition q is created which can describe the objects and only those objects in $ii(true)$.

Let $P = P \cup \{q\}$ and $P' = P' \cup \{q\}$. For each $q_l \in P' \setminus \{false\}$, let $v(q_l) = ii(q_l)$. Then, $v(q_1), \ldots, v(q_n)$, form a partition of U, because $v(q_l) \cap v(q_j) = \emptyset$ when $l \neq j$ and $v(q_l) \neq \emptyset$.

If we further define $W_l = v(q_l)$, for each $q_l \in P'$, then the set $\{W_1, \ldots, W_n\}$ specifies a collection of equivalence classes of a particular elementary equivalence relation R. Therefore, based on Definitions 10.5 and 10.8, structure

$$(U, R, P, P' \setminus \{false\}, val)$$

is a partial rough logic theory.

QED

However, in most cases, ii is defined on randomly derived formulae in $\mathcal{L}(P)$ rather than just on atomic propositions. For example, if $ii(windy) = U_1$, $ii(windy \wedge rain) = U_2$, and $ii(true) = U \setminus (U_1 \cup U_2)$, then this basic incidence assignment cannot produce an equivalence relation on U.

As a consequence, in most cases, it is not practical to construct rough logic theories from incidence calculus theories. Generally speaking, the amount of information sufficient to define an incidence calculus theory is significantly lower than that required to define rough logic theories, even a partial one.

10.2 Related Works

10.2.1 Interval Structures

An interval structure (see, e.g., [Wong et al, 1992], [Wong and Nie 1993], [Yao and Li, 1993], [Wong, 1995]) is a general purpose framework for representing uncertain information.

And, formally:

Definition 10.9: *Interval Structure*([Wong et al, 1992])

Given two mappings $\underline{F} : 2^U \rightarrow 2^{U'}$ *and* $\overline{F} : 2^U \rightarrow 2^{U'}$, *where* U *and* U' *are finite sets of objects, and if:*

- $\underline{F}$ *satisfies the axioms, for any subsets* $A, B \in 2^U$*:*

$$\begin{array}{ll} (L1) & \underline{F}(A \cup B) \supseteq \underline{F}(A) \cup \underline{F}(B), \\ (L2) & \underline{F}(A \cap B) = \underline{F}(A) \cap \underline{F}(B), \\ (L3) & \underline{F}(\emptyset) = \emptyset, \\ (L4) & \underline{F}(U) = U', \end{array}$$

- $\overline{F}$ *satisfies the axioms,for any subsets* $A, B \in 2^U$*:*

$$\begin{array}{ll} (U1) & \overline{F}(A \cup B) = \overline{F}(A) \cup \overline{F}(B), \\ (U2) & \overline{F}(A \cap B) \subseteq \overline{F}(A) \cap \overline{F}(B), \\ (U3) & \overline{F}(\emptyset) = \emptyset \\ (U4) & \overline{F}(U) = U' \end{array}$$

- *and, moreover,* $\overline{F}(A) = U' \setminus \underline{F}(\neg A)$*, where* $\neg A = U \setminus A$ *denotes the complement of* A*,*

then the pair $F = (\underline{F}, \overline{F})$ *is called an interval structure.*

Clearly, given an original incidence function, it is always possible to define an interval structure using the lower and upper bounds of incidences, (inf, sup), based on (2.1) and (2.2). Example 2.3 also demonstrates this equivalence relationship.

The Legal Assignment Finder, as well as the procedure of finding tighter bounds are no longer applicable in generalized incidence calculus because it is not truth functional anymore.

Therefore, it is not possible to compare these mechanisms for getting the tighter bounds between generalized incidence calculus and the interval structure.

However, function j_F – a *basic set assignment* in the interval structure – is semantically equivalent to the basic incidence assignment ii.

A basic set assignment $j_F : 2^U \rightarrow 2^{U'}$ satisfies the following three conditions, which are also possessed by a basic incidence assignment:

$$\begin{array}{ll} (A1) & j_F(\emptyset) = \emptyset, \\ (A2) & \bigcup_{A \subseteq U} = U', \\ (A3) & A \neq B \Rightarrow (j_F(A) \cap j_F(B)) = \emptyset, \end{array}$$

for any $A, B \in 2^U$.

j_F is only definable when a pair of lower and upper mappings forms an interval structure.

Assignment ii does not need this requirement because it always comes along with a generalized incidence function. Lower and upper bounds of incidences emerge from this single assignment.

In [Wong et al, 1992], the authors stated that

> "...the lower and upper approximates of the rough-set model, the lower and upper bounds in incidence calculus, and the belief and plausibility functions all obey the axioms of an interval structure".

This echoes the result from [Skowron and Grzymala-Busse, 1994] as well as our research results in presented Chapter 7 and Section 10.1 about the relationships between incidence calculus and DS theory and rough sets theory, respectively.

10.2.2 Bacchus's Propositional Probability Structure

Bacchus [Bacchus, 1990] suggested a unified structure in his language **Lp**:

> "**Propositional probability structures**: we define the following structure which we use to interpret the formulae of our language for propositional probabilities:
>
> $$M =< \mathcal{O}, S, v, \mu >$$
>
> where:
>
> - $\mathcal{O}$ is a set of individuals representing objects of the domain that one wishes to describe in the logic. $\mathcal{O}$ corresponds to the domain of discourse in the ordinary usage of first-order logic.
> - S is a set of possible worlds.
> - v is a function that associates an interpretation of the language with each world.
> - μ is a discrete probability function on S. That is, μ is a function that maps the elements of S to the real interval [0,1] such that $\Sigma_{s \in S} \mu(s) = 1$."

The interpretation of the truth value of a formula is explained as follows [Bacchus, 1990]:

> "In sum, the truth value assigned to a formula will depend on three items: the semantic structure or model M (which determines the probability distribution μ, the interpretation function v, and the domain of objects $\mathcal{O}$); the current world s; and the variable assignment function v. We now give the inductive specification of the truth assignment, writing $(M, s, v) \models \alpha$ if the formula α is assigned a truth value **true** by the triple and writing $t^{(M,v)}$ for the individual denoted by the term t in the triple.
>
> For every formula α, the term created by the probability operator $prob(\alpha)$ is given the interpretation:
>
> $$(prob(\alpha))^{(M,v)} = \mu\{s' \in S : (M, s', v) \models \alpha\}$$
>
> ..."

So, the probability of a formula is interpreted as the probability of the set of possible worlds which satisfy that formula.

The propositional probability structure given here is very similar to the incidence calculus theory structure except for that in an incidence calculus theory, a set of formulae (axioms) is particularly specified in $\mathcal{A}$.

Both structures have used the probability of a set of possible worlds to interpret the probability of a formula.

Although the two structures are similar in their appearance, nevertheless, there is a significant difference in their probability propagation procedures.

In incidence calculus, possible worlds remain to be the main material in the propagation of probabilities, that is, the probability of a formula is calculated through its incidence set $i(\phi)$.

In Bacchus's structure, this seems not to be the case. In other words, in [Bacchus, 1990] possible worlds are used to represent information, and their functions in further evidence propagation are not clear.

An advantage of Bacchus's structure is that a framework for representing first order logic has been defined. This part of work in incidence calculus remains to be done.

10.2.3 Multiple-valued Logics

Although both multiple-valued logics and incidence calculus are all about extending propositional logic from just two values to multiple values, the nature of these extensions are fundamentally different.

In [Malinowski, 1993], a three valued logic was created to extend traditional two valued propositional logic. An intermediate truth value *unknown* is introduced to identify those formulae for which the truth value is not decidable for the time being. A corresponding truth table (Chapter 2) is then established to work out the truth value of a compound formula when its sub-formulae are three valued. From this point of view, a three-valued logic system can represent a situation involving complete ignorance (unknown) (not partial ignorance) or absolute lack of information.

On the other hand, however, incidence calculus is not designed to represent ignorance in the first instance. Rather, it is a mechanism to represent whatever is known so far (information collected), a mechanism designated for a situation when uncertainty and incompleteness may be involved. The truth value of a formula is in the form of a degree of probability, not just simple 'yes', 'no', or 'unknown'. The formulae with *unknown* values are not explicitly identified, and "degrees of uncertainty should not be taken as intermediary truth values" [Dubois and Prade, 1994a].

Therefore, incidence calculus should not be regarded as a different format of multiple-valued logic.

10.3 Summary

The overall objective of this book has been to demonstrate the importance of coupling numerical calculi for reasoning under uncertainty with pure symbolic approaches (e.g., classical logics, default logic, and ATMS) in the context of representing and reasoning about knowledge and information. Under the guideline of this objective, only the approaches to combining two different mechanisms were emphasised, the actual mechanisms involved have not been introduced in detail (as, e.g., default logic).

The overview of research results given in Chapter 1 provides the reader with a fairly good picture (although maybe not complete) of a variety of integrated approaches/mechanisms that have been created in order to achieve the above objective in the past two decades.

Among a number of integrated frameworks, we have chosen incidence calculus as our main discussion theory in order to reveal the inevitable nature of these research work. There are two reasons for this choice: first of all, incidence calculus itself can perform both symbolic and numerical reasoning, secondly, this theory has close connections with both pure numerical reasoning calculi and symbolic inference systems.

10.3.1 Coupling Incidence Calculus with other Theories in Practice

Original incidence calculus sets a platform for performing inference symbolically and expressing numerically degrees of beliefs in statements. Although this platform suffers from a number of limitations, nevertheless, its attempt to bring two different streams of reasoning patterns into one mechanism deserves to be acknowledged.

Generalized incidence calculus has overcome the three drawbacks implied in the original incidence calculus, i.e., a limited ability to represent information, a lack of an efficient algorithm for recovering incidence assignments from numerical ones, and an inability to deal with multiple pieces of evidence from different sources.

However, truth functionality possessed by the original incidence calculus is no longer valid in the generalized incidence calculus. Therefore, the compositionality property in classical logics cannot be applied to calculate incidences of formulae in generalized incidence calculus.

On the other hand, the three extensions make it possible for generalized incidence calculus to be comparable with either a pure numerical calculus, (as, e.g., DS theory), or a symbolic one (as, e.g., the ATMS). The exploration of relationships between generalized incidence calculus and DS theory should address not only what these two theories have in common, but also what more can be achieved if they are both employed in one system to deal with information of a different nature. For instance, if a piece of evidence is given in a numerical way, it is better to use DS theory to represent and to propagate

Table 10.4: Situations of two players holding different cards

U	Player 1	Player 2
w_1	X	Y
w_2	X	Z
w_3	Y	X
w_4	Y	Z
w_5	Z	X
w_6	Z	Y

it. However, when two pieces of evidence are derived from the same source, it is better to use generalized incidence calculus to deal with them, even this process may involve the transformation of mass functions into incidence calculus theories in the first step and then vice versa.

Similarly, the study of relationships between incidence calculus and the ATMS may serve as a basis for the calculation of degrees of belief on sets of assumptions.

10.3.2 Where Incidences Come From?

When applying generalized incidence calculus to practical problems, one major difficulty is how to obtain those incidences. Although Algorithm B (from numerical assignments to incidence functions) shown in Chapter 4 provides an efficient procedure of recovering an incidence function from an initial numerical assignment, the underlying condition for the use of this algorithm is rather critical: the numerical assignment has to be closed under $\wedge$ on the given set of axioms. In many cases, this condition can hardly be satisfied.

Recent studies on reasoning about knowledge using rough sets reported in [Wong, 1998], and on relationships between rough sets and incidence calculus shown in the first half of this chapter, provide a possibility of defining incidences from a data table (sometimes called an information system or knowledge system [Pawlak, 1991]) through equivalence relations. For example, a data table given below (Table 10.4) is used to represent information about which card each of the two players (named 1 and 2) is holding, among three possible choices X, Y and Z ([Wong, 1998]). w_1 to w_6 are 6 possible worlds in which the situations of the two players holding different cards are supported.

Two elementary equivalence relations are derived based on this table:

$$U/R1 = \{\{w_1, w_2\}_{1X}, \{w_3, w_4\}_{1Y}, \{w_5, w_6\}_{1Z}\},$$

and

$$U/R2 = \{\{w_3, w_5\}_{2X}, \{w_1, w_6\}_{2Y}, \{w_2, w_4\}_{2Z}\},$$

where $\{w_1, w_2\}_{1X}$ indicates in which possible worlds player 1 holds card labelled X, etc.

Based on these two elementary equivalence relations, we can obtain the following incidence assignments:

$$i(1X) = \{w_1, w_2\}, \quad i(1Y) = \{w_3, w_4\}, \quad i(1Z) = \{w_5, w_6\},$$

and

$$i(2X) = \{w_3, w_5\}, \quad i(2Y) = \{w_1, w_6\}, \quad i(2Z) = \{w_2, w_4\}.$$

$1X, 1Y, 1Z, 2X, 2Y$, and $2Z$ are six atomic propositions in a set of atomic propositions P, where $1X$ stands for 'player one is holding card X', etc.

10.3.3 Significance of Numerical-symbolic Reasoning

Numerical uncertainty reasoning techniques will still be many people's favourite choice to represent uncertainty and beliefs. However, this overwhelming preference has not blocked some persistent efforts of seeking alternatives to reasoning under uncertainty in symbolic forms that may be exemplified by such works as [Liu and Wellman, 1998b] and [Castel *et al.*, 1998]. Situations where symbolic representation and reasoning is still in demand (use) will also be a realistic aspect of knowledge representation and reasoning.

Therefore, it is not too optimistic to say that in the future, research on symbolic reasoning, as well as the integration of symbolic and numerical reasoning will still be active, despite a dominant position of Bayesian belief networks.

Finally, we quote the following remarks from [Krause and Clark, 1993] to conclude the book:

> "... Our own view is that the next decade will see a further heightening of the importance of symbolic relations between quantitative data structures on the one hand, and increasing axiomatic specification of symbolic systems on the other.
>
> Future trends: the convergence of symbolic and quantitative methods?"

Bibliography

[Anrig and Monney, 1999] Anrig, B. and Monney, P.A., Using propositional logic to compute probabilities in multistate systems. *Int. J. of Approximate Reasoning* **20**, 113-143.

[Bacchus, 1988] Bacchus,F., *Representing and reasoning with probabilistic knowledge.* PhD thesis, The University of Alberta, 1988. Also available from University of Waterloo, Ontario, Canada N2L 3G1, 1-135.

[Bacchus, 1990] Bacchus, F., *Representing and reasoning with probabilistic knowledge.* The MIT Press, Boston.

[Bacchus *et al.*, 1994] Bacchus, F., Grove, A., Halpern, J. and Koller, D., Generating new beliefs from old. *Proceedings of the Tenth Conference on Uncertainty in Artificial Intelligence.* Lopez De Mantaras, R. and Poole, D. (eds.), Morgan Kaufmann, San Francisco, 37-45.

[Bacchus *et al.*, 1996] Bacchus, F., Grove, A., Halpern, J. and Koller, D., From statistical knowledge bases to degrees of beliefs. *Artificial Intelligence* **87**, 75-143.

[Baldwin, 1987] Baldwin,J.F., Evidential Support Logic Programming. *Fuzzy Sets and Systems* **24**, 1-26.

[Barnett, 1981] Barnett,J.A., Computational methods for a mathematical theory of evidence. *IJCAI-81*, Hayes, P. J. (ed.), William Kaufmann, 868-875.

[Benferhat *et al.*, 1995] Benferhat, S., Saffiotti, A. and Smets, Ph., Belief functions and default reasoning. *Proceedings of the Eleventh Conference on Uncertainty in Artificial Intelligence.* Besnard, P. and Hanks, S. (eds.), Morgan Kaufmann, San Francisco, 19-26

[Benferhat and Sossai, 1998] Benferhat, S. and Sossai, C., Merging Uncertain Knowledge Bases in a Possibilistic Logic Framework. *Proceedings of the Foorteenth Conference on Uncertainty in Artificial Intelligence.* Cooper, G. and Moral, s. (eds.), Morgan Kaufmann, San Francisco.

[Bhatnagar and Kanal, 1986] Bhatnagar, R.K. and Kanal,L.N., Handling uncertainty information: a review of numerical and non-numerical methods. In *Proceedings of the Second Conference on Uncertainty in Artificial Intelligence*, Kanal and Lemmer (eds), Elsevier, Amsterdam, 3-26.

[Black,1987] Black,P., Is Shafer general Bayes? *Proc. of Third AAAI Uncertainty in Artificial Intelligence workshop*, 2-9.

[Boldrin and Sossai, 1995] Boldrin, L. and Sossai, C., An algebraic semantics for probabilistic reasoning. *Proceedings of the Eleventh Conference on Uncertainty in Artificial Intelligence.* Besnard, P. and Hanks, S. (eds.), Morgan Kaufmann, San Francisco, 27-35.

[Bonissone and Tong 1985] Bonissone,P.P. and Tong,R.M., Reasoning with uncertainty in expert systems. *Int. J. Man-Machine Studies* **22**, 241-250.

[Bosc and Prade, 1993] Bosc, P. and Prade, H., An introduction to fuzzy set and possibility theory based approaches to the treatment of uncertainty and imprecision in database management systems: from needs to solutions, Catalina, California.

[Bundy, 1985] Bundy, A., Incidence calculus: a mechanism for probabilistic reasoning. *J. Automated Reasoning* **1**, 263-283.

[Bundy, 1986] Bundy, A., Correctness criteria of some algorithms for uncertain reasoning using incidence calculus. *J. Automated Reasoning* **2**, 109-126.

[Bundy, 1992] Bundy, A., Incidence calculus. *The Encyclopedia of Artificial Intelligence*, 663-668. Also available as Research Paper 497, Dept. of Artificial Intelligence, Edinburgh.

[Carnap, 1962] Carnap, R., *Logical foundations of probabilities.* University of Chicago Press, Chicago.

[Castel *et al.*, 1998] Castel, C., Cossart, C. and Tessier, C., Dealing with uncertainty in situation assessment: towards a symbolic approach. *Proceedings of the Fourteenth Conference on Uncertainty in Artificial Intelligence.* Cooper, G. and Moral, S. (eds.), Morgan Kaufmann, San Francisco.

[Chateauneuf, 1994] Chateauneuf, A., Combination of compatible belief functions and relation of specificity. In *Advances in the Dempster-Shafer theory of evidence*, Yager, Fedrizzi and Kacprzyk (eds.), Wiley, New York, 97-114.

[Chang and Fung, 1991a] CHang, K.C. and Fung, R., Symbolic probabilistic inference with evidence potential. *Proceedings of the Seventh Conference on Uncertainty in Artificial Intelligence.* D'Ambrosio, B.D., Smets, Ph., Bonissone, P.P. (eds.), Morgan Kaufmann, San Mateo, California, 82-85.

[Chang and Fung, 1991b] CHang, K.C. and Fung, R., Symbolic probabilistic inference with continuous variables. *Proceedings of the Seventh Conference on Uncertainty in Artificial Intelligence.* D'Ambrosio, B.D., Smets, Ph., Bonissone, P.P. (eds.), Morgan Kaufmann, San Mateo, California, 77-81.

[Cloteaux, et al, 1998] Cloteaux, B., Eick, C., Bouchon-Meunier, B. and Kreinovich, V., From ordered beliefs to numbers: how to elicit numbers without asking for them. *Int. J. Intelligent Systems* **13**, 801-820.

[Cohen, 1985] Pohen, P.R., *Heuristic Reasoning about Uncertainty: an artificial intelligence approach.* PhD Dissertation. Pitman Advanced Publishing Program, Boston, London, Melbourne.

[Corlett and Todd, 1985] Corlett, R.A. and Todd, S.J., A Monte-Carlo approach to uncertain inference. *Proceedings of AISB-85*, Ross, P. ed., 28-34.

[Correa da Silva and Bundy, 1990] Correa da Silva,F.S. and Bundy,A., On some equivalence relations between incidence calculus and Dempster-Shafer theory of evidence. *Proc. of the sixth conference on uncertainty in artificial intelligence.* Bonissone, P.P., Henrion, M., Kanal, L.N. and Lemmer, J.F. (eds), Elsevier, New York, 378-383,

[Correa da Silva and Bundy, 1991] Correa da Silva,F.S. and Bundy,A., A rational reconstruction of incidence calculus. Research Paper 517. Dept. of Artificial Intelligence, University of Edinburgh.

[d'Ambrosio, 1988] d'Ambrosio,B., A hybird approach to reasoning under uncertainty. *Int. J. Approx. Reasoning* **2**, 29-45.

[d'Ambrosio, 1990] d'Ambrosio,B., Incremental construction and evaluation of defeasible probabilistic models. *Int. J. Approx. Reasoning* **4**, 233-260.

[Darwiche and Goldszmidt, 1994] Darwiche, A. and Goldszmidt, M., On the relation between kappa calculus and probabilistic reasoning. *Proceedings of the Tenth Conference on Uncertainty in Artificial Intelligence.* Lopez de Mantaras, R. and Poole, D. (eds.), Morgan Kaufmann Publishers, San Francisco, California, 145-153.

[de Kleer, 1986a] de Kleer,J., An assumption-based TMS. *Artificial Intelligence* **28**, :127-162.

[de Kleer, 1986b] de Kleer, J., Extending the ATMS. *Artificial Intelligence* **28**, 163-196.

[de Kleer and Williams, 1987] de Kleer, J. and Williams, B.C., Diagnosing multiple faults. *Artificial Intelligence* **32**, 97-130.

[Dempster, 1967] Dempster, A.P., Upper and lower probabilities induced by a multivalued mapping. *Ann. Math. Stat.* **38**, 325-339.

[Delgrande, 1988] Delgrande, J.P., An approach to default reasoning based on first order conditional logic: revised report. *Artificial Intelligence* **36**, 63-90.

[Doyle, 1979] Doyle,J., A truth maintenance system. *Artificial Intelligence* **12**, 231-272.

[Druzdzel and Henrion, 1993] Druzdzel, M.J. and Henrion, M., Efficient reasoning in qualitative probabilistic networks. *Proceedings of the Eleventh National Conference on Artificial Intelligence*, 548-553.

[Dubois and Prade,1986] Dubois,D. and Prade, H., On the unicity of Dempster rule of combination. *Int. J. Intelligent Systems* **1** ,133-142.

[Dubois and Prade 1987] Dubois, D. and Prade, H., Necessity measures and the resolution principle. *IEEE Trans. Systems, Man, and Cybernetics* **17**, 474-478.

[Dubois and Prade,1988] Dubois,D. and Prade, H., An introduction to possibilistic and fuzzy logics. In *Non-standard logics for automated reasoning* Smets, Ph., Mamdani, E.H., Dubois, D. and Prade, H. (eds.), Academic Press, New York, 287-326.

[Dubois and Prade,1988a] Dubois,D. and Prade, H., *Possibility Theory: An Approach to the Computerized processing of Uncertainty.* Plenum Press, New York.

[Dubois and Prade 1990a] Dubois, D. and Prade, H., Consonant approximations of belief functions. *Int. J. Approx. Reasoning* **4**, 419-449.

[Dubois and Prade 1990b] Dubois, D. and Prade, H., Resolution principles in possibilistic logic. *Int. J. Approximate Reasoning* **4**, 1-21.

[Dubois and Prade 1990c] Dubois, D. and Prade, H., Updating with belief functions, ordinal conditional functions and possibility measures. *Proc. of the Sixth Conference on Uncertainty in Artificial Intelligence.* Bonissone, P.P., Henrion, M., Kanal, L.N. and Lemmer, J.F. (eds), Elsevier, New York, , 419-449.

[Dubois *et al.*, 1987] Dubois, D., Lang, J. and Prade, H., Theorem proving under uncertainty – A possibility theory based approach. *Proceedings of 10th Int. Joint Conference on Artificial Intelligence,* McDermott, J.P. (ed.), Morgan Kaufmann, San Mateo, 984-986.

[Dubois *et al.*, 1989] Dubois, D., Lang, J. and Prade, H., Automated reasoning using possibilistic logic: semantics, belief revision and variable certainty weights. *Proceedings of Fifth Conference on Uncertainty in Artificial Intelligence,* Henrion, M., Shachter, R.D., Kanal, L.N., and Lemmer, J.F. (eds.), Elsevier, New York, 81-87.

[Dubois *et al.*, 1990] Dubois, D., Lang, J. and Prade, H., Handling uncertain knowledge in an ATMS using possibilistic logic. *ECAI-90 Workshop on Truth Maintenance Systems.* Stockholm, Sweden, 87-106.

[Dubois and Prade, 1992a] Dubois,D. and Prade, H., Belief change and possibility theory. In *Belief Revision* P.Gardenfors (ed.), Cambridge University Press, 142-182.

[Dubois and Prade 1992b] Dubois, D. and Prade, H., Evidence, knowledge and belief functions. *Int. J. Approx. Reasoning* **6**, 295-319.

[Dubois and Prade, 1994a] Dubois, D. and Prade, H., Non-standard theories of uncertainty in knowledge representation and reasoning. *The Knowledge Engineering Review,* **9**(4), 399-416.

[Dubois and Prade, 1994b] Dubois, D., Lang, J. and Prade, H., Possibilistic logic. In *Handbook of Logic in Artificial Intelligence and Logic Programming* vol 3. Gannay, D., Hogger, C. and Robinson, J. (eds.), Clarendon Press, 439-513.

[Duda *et al.*, 1976] Duda,R.O., Hart, P.E. and Nilsson, N., Subjective Bayesian methods for rule-base inference systems. *Proceedings 1976 National Computer Conference, AFIPS* **45**, 1075-1082.

[Düntsch, 1997] Düntsch, I., A logic for rough stes. *Theoretical Computer Science* **179**(1-2), 427-236.

[Düntsch and Gediga, 1999] Düntsch, I. and Gediga, G., Rough set data analysis. to appear in *Encyclopedia of Computer Science and Technology.*

[Frisch and Haddawy, 1994] Frisch, A. and Haddawy, P., Anytime deduction for probabilistic logic. *Artificial Intelligence* **69**, 93-122.

[Fagin *et al.*, 1988] Fagin, R., Halpern, J.Y. and Megiddo, N., A logic for reasoning about probabilities. technical Report RJ 6190 4/88, IBM Research, Almaden Research Center, 650 Harry Road, San Jose, California, 95120-6099.

[Fagin and Halpern, 1989a] Fagin,R. and Halpern, J.Y., Uncertainty, belief and probability. **IJCAI-89**, Sridharan, N.S. (ed.), Morgan Kaufmann, San Mateo, 1161-1167.

[Fagin and Halpern, 1989b] Fagin,R. and Halpern, J.Y., Uncertainty, belief and probability. Research Report of IBM, RJ 6191.

[Fagin and Halpern, 1990] Fagin,R. and Halpern, J.Y., A new appraoch to updating beliefs. *Proceedings of the Sixth Conference on Uncertainty in Artificial Intelligence.* Bonissone, P.P., Henrion, M., Kanal, L.N. and Lemmer, J.F. (eds), Elsevier, New York, 347-374.

[Fagin *et al.*, 1995] Fagin, R., Halpern, J.Y., Moses, Y. and Vardi, M., *Reasoning about Knowledge.* The MIT Press. Cambridge, Massachusetts.

[Field, 1977] Field, H., Logic, meaning, and conceptual role. *J. Philosophy* **77**, 374-409.

[Fringuelli *et al.*, 1991] Fringuelli, B., Marcugini, S., Milani, A. and Rivoira, S., A reason maintenancs system dealing with vague data. *Proceedings of Seventh Uncertainty in Artificial Intelligence.* D'Ambrosio, B, D., Smets, Ph. and Bonissone, P. P., Morgan Kaufmann, San Mateo, 111-117.

[Fox, et al, 1993] Fox, J., Krause, P. and Elvang-Goransson, M., Argumentation as a general framework for uncertain reasoning. *Proceedings of the Ninth Conference on Uncertainty in Artificial Intelligence.* Heckerman, D. and Mamdani, E.H. (eds.), Stanford, Morgan Kaufmann, San Mateo, 428-434,

[Fulvio Monai and Chehire, 1992] Fulvio Monai,F. and Chehire, T., Possibilistic assumption based truth maintenance systems, validation in a data fusion application. *Proceedings of the Eighth Conference on Uncertainty in Artificial Intelligence.* Dubois, D., Wellman, M.P., D'Ambrosio,B. and Smets, Ph. (eds.), Morgan Kaufmann, San Mateo, 83-91,

[Gardenfors, 1992] Gardenfors,P., Belief revision: an introduction. In *Belief Revision*, Gardenfors (ed.), Cambridge University Press, 1-28.

[Ginsberg, 1984] Ginsberg, M.L., Non-monotonic reasoning using Dempster's rule. *Proceedings of AAAI-84*, 126-129.

[Goldszmidt and Pearl, 1992] Goldszmidt, M. and Pearl, J., Stratified rankings for causal relations. *Proceedings of the Fourth International Workshop on Non-monotonoc Reasoning*, Vermont, 99-110.

[Goldszmidt and Pearl, 1996] Goldszmidt, M. and Pearl, J., Qualitative probabilities for default reasoning, belief revision and causal modeling. *Artificial Intelligence* **84**, 57-112.

[Gordon and Shortliffe, 1985] Gordon,J. and Shortliffe, E.H., A method for managing evidential reasoning in a hierarchical hypothesis space. *Artificial Intelligence* **26**, 323-357.

[Grosof, 1986] Grosof, B.N., An inequality paradigm for probabilistic knowledge: the logiv vonditional probability intervals. *Proceedings of the Second Conference on Uncertainty in Artificial Intelligence.* Kanal, L.N. and Lemmer, J.F. (eds.), Elsevier, New York, 259-275.

[Guan and Bell, 1991] Guan, J.W. and Bell, D.A., *Evidential reasoning and its applications* Vol 1, North-Holland, Amsterdam.

[Halpern, 1990] Halpern, J.Y., An analysis of first order logics of probability. *Artificial Intelligence* **46**, 311-350.

[Halpern and Fagin, 1992] Halpern, J.Y. and Fagin, R., Two views of belief: belief as generalized probability and belief as evidence. *Artificial Intelligence* **54**, 275-317.

[Hau and Kashyap, 1990] Hau, H.Y. and Kashyap, R.L., Belief combination and propagation in a lattice-structured inference network. *IEEE Trans. Syst., Man and Cybern.* **20**(1), 45-57.

[Heinsohn, 1991] Heinsohn, J., A hybird approach for modelling uncertainty in terminological logics. *Proceedings of European Conference on Symbolic and Quantitative Approaches to Uncertainty*, Marseilles, 198-205.

[Heinsohn, 1994] Heinsohn, J., Probabilistic description logics. *Proceedings of the Tenth Conference on Uncertainty in Artificial Intelligence.* Lopez de Mantaras, R. and Poole, D. (eds.), Morgan Kaufmann, San Francisco, California, 311-318.

[Henrion and Druzdzel, 1990] Henrion, M. and Druzdzel, M.J., Quanlitative propagation and scenario based schemes for explaining probabilistic reasoning. *Proceedings of Sixth Conference Uncertainty in Artificial Intelligence.* Bonissone, P.P., Henrion, M., Kanal, L.N. and Lemmer, J.F., (eds.), Elsevier, New York, 17-32.

[Henrion *et al.*, 1994] Henrion, M., Proven, G., Del Favero, B., and Sanders, G., An experimental comparison of numerical and quanlitative probabilistic reasoning. *Proceedings of the Tenth Conference on Uncertainty in Artificial Intelligence.* Lopez de Mantaras, R. and Poole, D. (eds.), Morgan Kaufmann, San Francisco, California, 319-327.

[Hong *et al.*, 1999] Hong, X., Adamson, K. and Liu, W., Using parallel technique to improve the computational efficiency of evidential combination. *Proceedings of Tools on Artificial Intelligence*, November, Chicago, Illinois, 55-58.

[Hong, 2001] Hong, X., Reasoning under Uncertainty and Parallelism for Medical Diagnosis. *PhD thesis*, to appear.

[Hunter,1987] Hunter,D., Dempster-Shafer vs. probabilistic logic. *Proc. Third AAAI uncertainty in artificial intelligence workshop*, Kanal, L. N., Levitt, T.S. and Lemmer, J.F. (eds.), North-Holland, Amsterdam, 22-29.

[Jeffrey, 1965] Jeffrey, R., *The logic of decisions.* McGraw-Hill, New York.

[Jensen *et al.*, 1990] Jensen, F.V., Lauritzen, S.L. and Olesen, K.G., Bayesian updating in causal probabilistic networks by local computation. *Comput. Statist. Quart.* **4**, 269-282.

[Kennes, 1991] Kennes,R., Evidential reasoning in a categorical perspective: conjunction and disjunction of belief functions. *Proc. of the Seventh Conference on Uncertainty in Artificial Intelligence.* D'Ambrosio, B.D., Smets, Ph. and Bonissone, P.P. (eds.), Morgan Kaufmann, San Mateo, 174-181.

[Kennes and Smets, 1990] Kennes,R. and Smets,Ph., Computational aspects of the Mobius transform. *Procs. of the Sixth Conference on Uncertainty in Artificial Intelligence.* Bonissone, P.P., Herionn, M., Kanal and Lemmer (eds.), North-Holland, Amsterdam, 401-416.

[Kennes, 1992] Kennes,R., Computational aspects of the Mobius transform of a graph. *IEEE Trans. Syst., Man Cybern.* **22**, 201-223.

[Kohlas and Monney, 1993] Kohlas,J. and Monney, P.A., Probabilistic assumption-based reasoning. *Proc. of Ninth Conference on Uncertainty in Artificial Intelligence.* Heckermann, D. and Mamdani, E.H. (eds.), Morgan Kaufmann, San Mateo, 485-491.

[Kohlas *et al.*, 1998] Kohlas,J., Anrig, B., Haenni, R. and Monney, P.A., Model-based diagnostic and probabilistic assumption-based reasoning. *Artificial Intelligence* **104**, 71-106.

[Komorowski *et al.*, 1999] Komorowski, J., Polkowski, L. and Skowron, A., Rough sets: a tutorial. Unpublished manuscript, Poland.

[Kong, 1987] Kong,A., *Multivariate belief functions and graphical methods.* PhD thesis, Dept. of Statistics, Harvard University.

[Kruse *et al.*, 1992] Kruse,R., Schwecke, E. and Heinsohn, J., *Uncertainty and Vagueness in Knowledge-Based Systems.* Springer-Verlag, Berlin.

[Krause and Clark, 1993] Krause, P. and Clark, D., *Representing Uncertain Knowledge: An Artificial Intelligence Approach.* Kluwer, Dordrecht.

[Kyburg, 1987] Kyburg Jr., H.E., Bayesian and non-Bayesian evidential updating. *Artificial Intelligence* **31**, 271-293.

[Laskey and Lehner, 1989] Laskey, K.B. and Lehner, P.E., Assumptions, beliefs and probabilities. *Artificial Intelligence* **41**, 65-77.

[Lauritzen and Spiegelhalter, 1988] Lauritzen , S.L. and Spiegelhalter, D.J., Fast manipulation of probabilities with local representation-with applications to expert systems. *J. Royal Statistical Society B***50** (**2**),157-224.

[LeBlanc, 1983] LeBlanc, H., Alternatives to standard first-order semantics. In *Handbooks of Philosophical Logic. Vol II.* Gabbay, D. and Guenthner, F. (eds.), D. Reidel, Dordrecht, 225-258.

[Lemmer, 1986] Lemmer, J.F., Confidence factors, empiricism and Dempster-Shafer theory of evidence, *Proc. of the Second Conference on Uncertainty in Artificial Intelligence.* Kanal, L.N. and Lemmer, J.F. (eds), Elsevier, New York, 117-125.

[Lin and Wildberger, 1995] Lin, T.Y. and Liu, Q., First order rough logic I, Approximate reasoning via rough sets. *Fundmentae Informaticae* **27**, 137-153.

[Lin and Wildberger, 1994] Lin, T.Y. and Wildberger, A.M. (eds), *Soft Computing. Proceedings of Third International Workshop on Rough Sets and Soft Computing.* San Jose State University, San Jose, California.

[Lingras and Wong, 1990] Lingras,P. and Wong, S.K.M., Two perspectives of the Dempster-Shafer theory of belief functions. *Int. J. Man-Machine Studies* **33**, 467-487.

[Liu et al, 1992] Liu, W., Hughes, J.G. and McTear, M.F., Representing heuristic knowledge in DS theory. *Proc. of the Eighth Conference on Uncertainty in Artificial Intelligence.* Dubois, D., Wellman, M., D'Ambrosio and Smets, Ph. (eds.), Morgan Kaufmann, San Mateo, 182-190,

[Liu *et al.*, 1993] Liu, W., Hong, J., McTear, M.F. and Hughes, J.G., An extended framework for evidential reasoning systems. *Int. J. Pattern Recognition and Artificial Intelligence* **7**(3), 441-457.

[Liu *et al.*, 1994] Liu, W., Hughes, J.G. and McTear, M.F., Representing heuristic knowledge and propagating beliefs in the Dempster-Shafer theory of evidence. In *Advances in the Dempster-Shafer Theory of Evidence.* Fedrzzi, Kacprzyk and Yager (eds), Wiley, New York, 441-472.

[Liu *et al.*, 1993a] Liu, W., Bundy, A. and Robertson, D., Recovering incidence functions. *Proceedings of 2nd European conference on symbolic and quantitative approaches to reasoning and uncertainty.* Lecture Notes in Computer Science 747, 241-248, Spinger-Verlag, Berlin (Long version is available as Department Research Paper 648. Department of Artificial Intelligence, University of Edinburgh).

[Liu *et al.*, 1993b] Liu, W., Bundy, A. and Robertson, D., On the relations between incidence calculus and ATMS. *Proceedings of 2nd European conference on symbolic and quantitative approaches to reasoning and uncertainty.* Lecture Notes in Computer Science 747, 248-256 Spinger-Verlag, Berlin (also available as Department Research Paper 611. Department of Artificial Intelligence, University of Edinburgh).

[Liu and Bundy, 1994] Liu, W. and Bundy, A., A comprehensive comparison between generalized incidence calculus and DS theory. *The Int. J. Human-Computer Studies* formerly *The Int. J. Man-Machine Studies* **40**, 1009-1032.

[Liu and Bundy, 1996] Liu, W. and Bundy, A., Construct probabilistic ATMS using extended incidence calculus. *Int. J. Approximate Reasoning* **Vol 15**, 145-182.

[Liu 1996] Liu, W., The incidence propagation method. *Proceedings of the 1996 International Symposium on Multiple-Valued Logic*, May, Spain, IEEE Computer Society, 124-129.

[Liu *et al.*, 1998] Liu, W., McBryan, D. and Bundy, A., The method of assigning incidences. *Int. J. of Applied Intelligence. In Correa da Silva (guest ed.): Special Issue on Systems for Uncertain Reasoning* **9** (2), 139-161. .

[Liu, 1998] Liu, W., A domain independent data structure for telecommunications using adapted ATMS. *Proceedings of 7th Int. Conf. on Info. Processing and Management of Uncertainty in Knowledge-Based Systems (IPMU'98)*, Paris, 1824-1829,

[Liu and Hong, 1999] Liu, W. and Hong, J., Re-investigating Dempster's idea on evidence combination. *Knowledge and Information Systems: An Int. J.*, **2** (2), May 2000.

[Liu and Wellman, 1998a] Liu, C.L. and Wellman, M.P., Incremental tradeoff resolution in qualitative probabilistic networks. *Proceedings of the Fourteenth Conference on Uncertainty in Artificial Intelligence.* Cooper, G.F. and Moral, F. (eds.), Morgan Kaufmann, San Francisco, 338-345.

[Liu and Wellman, 1998b] Liu, C.L. and Wellman, M.P., Using Qualitative Relationships for Bounding Probability Distributions *Proceedings of the Fourteenth Conference on Uncertainty in Artificial Intelligence.* Cooper, G.F. and Moral, F. (eds.), Morgan Kaufmann, San Francisco, 330-337.

[Liu, 1986] Liu, G.S.H., Causal and plausible reasoning in expert systems. *AAAI-86*, 220-225.

[Lowrance *et al.*, 1981] Lowrance, J.D., and Garvey, T.D., Evidential reasoning: an developing concept. *Proceedings IEEE International Conference on Cybernetics and Society*, 6-9.

[Lu and Stephanou, 1984] Lu, S.Y. and Stephanou, H.E., A set-theory framework for the processing of uncertain knowledge. *Proceedings of AAAI-84*, 216-221.

[Maier, 1983] Maier, D., *The Theory of Relational Database.* Computer Science Press.

[Maheshwari and Shen, 1998] Maheshwari, P. and Shen, H., An efficient clustering algorithm for partitioning parallel programs. *Parallel Computing* **24**, 893-909.

[Malinowski, 1993] Malinowski, G., *Many-valued logics.* Oxford Logic Quides:25. General Editors: D. Gabbay, A. Macintyre, and D. Scott. Oxford Science Publications.

[Mamdani *et al.*, 1988] Mamdani, A., Bigham, J. and Sheridan, F., Introduction. In *Non-Standard Logics for Automated Reasoning.* Smets, Mamdani, Dubois and Prade (eds), Academic Press, 1-25.

[McBryan, 1996] McBryan, D., *Assigning incidences to formulae.* Fourth year Project, Department of Artificial Intelligence, University of Edinburgh.

[McBurney and Parsons, 2000] McBurney, P. and Parsons, S., Risk agoras: dialectical argumentation for scientific reasoning. *Proceedings of the Sixteenth conference on Uncertainty in Artificial Intelligence.* Boutilier, C. and Goldszmidt, M. (eds), Morgan Kaufmann, San Francisco.

[McCarthy, 1980] McCarthy, J., Circumscription - a form of nonmonotonic reasoning. *Artificial Intelligence* **13**(1-2), 27-39.

[McDermott and Doyle, 1980] McDermott, D. and Doyle, J., Nonmonotonic logic I. *Artificial Intelligence* **13**(1-2), 41-72.

[McLean, 1992] McLean, D.R., *Testing and extending the incidence calculus,* MSc Dissertation, Department of Artificial Intelligence, University of Edinburgh.

[McLean *et al.*, 1995] McLean, D.R., Bundy, A. and Liu, W., Assignment methods for incidence calculus. *Int. J. Approximate Reasoning* **12**, 21-41 (also available as Department Research Paper 649. Department of Artificial Intelligence, University of Edinburgh).

[McLeish, 1988] McLeith, M., Probabilistic logic: some comments and possible use for nonmonotonic reasoning. In *Proceedings of the Fourth Conference on Uncertainty in Artificial Intelligence.* Lemmer, J.F. and Kanal, L.N. (eds.), Elsevier (North-Holland), Amsterdam, 55-62,

[McLeish, 1989] McLeith, M., Nilsson's probabilistic entailment extended to Dempater Shafer theory. *Proceedings of the Fifth Conference on Uncertainty in Artificial Intelligence.* Knal, L.N., Levitt, T.S. and Lemmer, J.F. (eds.), Elsevier (North-Holland), Amsterdam.

[McLeish, 1990] McLeith, M., A model for non-monotonic reasoning using Dempster's combination rule. *Proc. of the Sixth Conference on Uncertainty in Artificial Intelligence.* Bonissone, P. P.,Henrion, M., Kanal, L. N. and Lemmer, J. F. (eds), Elsevier, New York, 518-528.

[Mellouli, 1987] Mellouli, K., *On the propagation of beliefs in networks using the Dempster-Shafer theory of evidence.* PhD thesis.

[Nilsson, 1986] Nilsson, N.L., Probabilistic logic, *Artificial Intelligence* **28**, 71-87.

[Nguyen and Smets, 1993] Nguyen, H.T. and Smets, Ph., On Dynamics of Cautious Belief and Conditional Objects. *Int. J. Approx. Reasoning* **8**, 89-104.

[Orlowska, 1998] Orlowska, E., 1998 (ed.) *Incomplete Information: Rough Set Analysis.* Physica-Verlag (Springer-Verlag). In *Studies in Fuzziness and Soft Computing,* Volume 13, Kacprzyk, J. (series ed.).

[Parsons, 1995] Parsons, S., Refining reasoning in quanlitative probabilistic reasoning. *Proceedings of the Eleventh Conference on Uncertainty in Artificial Intelligence.* Besnard, P. and Hanks, S. (eds.), Morgan Kaufmann, San Francisco, 427-434.

[Parsons, 1996] Parsons, S., Current approaches to hanlding imperfect information in data and knowledge bases. *IEEE Trans. Knowledge and Data Engineering* **8**(3), 353-369.

[Pawlak, 1982] Pawlak, Z. Rough Sets. *Int. J. Computer and Information Sciences* **11**, 341-356.

[Pawlak *et al.*, 1988] Pawlak, Z., Wong, S.K.M. and Ziarko, W., Rough sets: probabilistic versus deterministic approaches. *Int. J. Man-Machine Studies* **29**, 81-95.

[Pawlak, 1991] Pawlak, Z., *Rough Sets. Theoretical Aspects of Reasoning about Data.* Kluwer, Dordrecht/Boston/London.

[Pearl, 1986] Pearl, J., Fusion, propagation, and structuring in Bayesian networks. *Artificial Intelligence* **29**, 241-288.

[Pearl, 1988] Pearl, J., *Probabilistic Reasoning in Intelligence Systems: networks of plausible inference.* Morgan Kaufmann, San Mateo.

[Pearl, 1992] Pearl, J., Rejoinder to Comments on "Reasoning with belief function: An analysis of compatibility". *Int. J. Approximate Reasoning* **6**, 425-443.

[Polkowski and Skowron, 1998a] Polkowski, L. and Skowron, A. (ed), *Rough Sets in Knowledge Discovery 1: methodology and applications.* Physica-Verlag (Springer-Verlag). In *Studies in Fuzziness and Soft Computing,* Vol 18, Kacprzyk, J. (series ed.).

[Polkowski and Skowron, 1998b] Polkowski, L. and Skowron, A. (ed), *Rough Sets in Knowledge Discovery 2: applications, case studies and software systems.* Physica-Verlag (Springer-Verlag). In *Studies in Fuzziness and Soft Computing,* Vol 19, Kacprzyk, J. (series ed.).

[Provan, 1989] Provan, G.M., An analysis of ATMS-based techniques for computing Dempster-Shafer belief functions. *Proc. of Eleventh International Joint Conference on Artificial Intelligence,* Sridharan, N. S. (ed.), Morgan Kaufmann, San Mateo, 1115-1120.

[Quillian, 1968] Quillian, M. R., Semantic memory. *Proceedings of Semantic Information,* Minsky, M. (ed.), MIT Press, Boston, 216-270.

[Reiter, 1980] Reiter, R., A logic for default reasoning. *Artificial Intelligence* **13**(1-2), 81-132.

[Renooij and van der Gaag, 1999] Renooij, S. and van der Gaag, L.C., Enhancing QPNs for Trade-off Resolution. *Proceedings of the Fifteenth Conference on Uncertainty in Artificial Intelligence.* Laskey, K.B. and Prade, H. (eds), Morgan Kaufmann, San Francisco.

[Renooij, et al, 2000] Renooij, S., van der Gaag, L.C., Parsons, S., and Green, S., Pivotal Pruning of Trade-offs in QPNs. *Proceedings of the Sixteenth Conference on Uncertainty in Artificial Intelligence.* Boutilier, C. and Goldszmidt, M. (eds), Morgan Kaufmann, San Francisco.

[Rich, 1983] Rich, E., Default reasoning as likelihood reasoning. *Proceedings of AAAI-83,* 348-351.

[Ruspini, 1987] Ruspini, E.H., Epistemic logics, probability, and the calculus of evidence. *Proceedings of the Tenth International Joint Conference on Artificial Intelligence.* McDermott, J. P. (ed.), Milan, Italy, August, Morgan Kaufmann, San Mateo, 924-931.

[Saffiotti, 1992] Saffiotti, A., A belief function logic. *Proceedings of the Tenth National Conference on Artificial Intelligence,* Boston, 642-647.

[Schaub, 1997] Schaub, T., *The Automation of Reasoning with Incomplete Information: From semantic foundations to efficient computation.* Lecture Notes in Artificial Intelligence 1049, Subseries of Lecture Notes in Computer Science. Springer-Verlag.

[Scott and Krauss, 1966] Scott, D. and Krauss, P., Assigning probabilities to logical formulas. In *Aspects of Inductive Logic*. Hintikka, J. and Suppes, P. (eds.), North-Holland, Amsterdam.

[Shachter *et al.*, 1990] Shachter, R., Del Favero, A. and D'Ambrosio, B., Symbolic probabilistic inference: a probabilistic perspective. *Proceedings of AAAI-90.*

[Shafer, 1976] Shafer, G., *A Mathematical theory of evidence*, Princeton University Press.

[Shafer, 1981] Shafer, G., Jeffrey's rule of conditioning. *Philisophy of Science* **48**, 337-362.

[Shafer, 1982] Shafer, G., Belief functions and parametric models. *J. Royal Statistical Society Series B* 44, 322-352.

[Shafer and Tversky, 1985] Shafer, G. and Tversky, A., Languages and Designs for Probability Judgement. *Cognitive Science* **9**, 309-339.

[Shafer, 1985] Shafer, G., Hierarchical evidence. *Proceedings of the 2nd conference on AI Applications*, Miami Beach, FL, IEEE Computer Society Press, 16-25.

[Shafer, 1986] Shafer, G., Probability judgement in artificial intelligence. *Proceedings of the Second Conference on Uncertainty in Artificial Intelligence.* Kanal, L.N. and Lemmer, J.F. (eds.), Elsevier, New York, 127-135.

[Shafer, 1987] Shafer, G., Probability judgement in artificial intelligence and expert systems. *Stat. Sci.* **2**, 3-16.

[Shafer 1987b] Shafer, G., Belief function and possibility measures. *Analysis of Fussy Information* **1**, J.C. Bezdek (ed.), CRC Press, Boca Raton, 51-84

[Shafer and Logan, 1987] Shafer, G. and Logan, R., Implementing Dempster's rule for hierarchical evidence. *Artificial Intelligence* **33**, 271-298.

[Shafer *et al.*, 1987] Shafer, G., Shenoy, P.P., and Mellouli, K., Propagating belief functions in qualitative Markvo trees. *Int. J. Approximate Reasoning* **1**,349-400.

[Shafer and Shenoy, 1988] Shafer, G. and Shenoy, P.P., Bayesian and belief function propagation. Working Paper No. 192, School of Business, University of Kansas.

[Shafer, 1990] Shafer, G., Perspectives on the theory and practice of belief functions. Working paper No.218, School of Business, University of Kansas.

[Shafer and Pearl, 1990] Shafer, G. and Pearl, J. (eds.), *Readings in Uncertain Reasoning.* Morgan Kaufmann, San Mateo.

[Shenoy and Shafer, 1986] Shenoy, P.P. and Shafer, G., Propagating belief functions with local computations. *IEEE Expert* **1**(3), 43-52.

[Shenoy and Shafer, 1990] Shenoy, P.P. and Shafer, G. , Axioms for probability and belief function propagation. In *Readings in Uncertain Reasoning.* Shafer and Pearl (eds.), Morgan Kaufmann, San Mateo, 575-610.

[Shenoy, 1997] Shenoy, P.P., Binary join trees for computing marhinals in the Shenoy-Shafer architecture. *Int. J. Approximate Reasoning* **17**, 239-263.

[Shortliffe, 1976] Shortliffe, E.H., **Computer-based medical consultations: MYCIN**. Elsevier, New York.

[Shrobe, 1988] Shrobe, H.E. (ed.), *Exploring Artificial Intelligence.* Survey talks from the National Conference on Artificial Intelligence in 1988.

[Skowron and Grzymala-Busse, 1994] Skowron, A. and Grzymala-Busse, J., From the rough set theory to evidence theory. In *Advances in the Dempster-Shafer theory of evidence.* Yager, R.R., Fedrizzi, M. and Kacprzyk, J. (eds.), Wiley, New York, 193-235.

[Slowinski, 1992] S"lowi"nski, R. (ed.), *Intelligent Decision Support: Handbook of Applications and Advances of the Rough Sets Theory.* Kluwer, Boston.

[Smets, 1988] Smets,Ph., Belief functions. In *Non-Standard Logics for Automated Reasoning.* Smets, Ph., Mamdani, E.H., Dubois, D. and Prade, H. (eds.), Academic Press, Harcourt Brace Jovanovich, 253-286.

[Smets, 1990] Smets, Ph. The Combination of Evidence in the Transferable Belief Model. *IEEE Trans. Pattern Analysis and Machine Intelligence* **12**(5), 447-458.

[Smets and Kennes, 1994] Smets, Ph. and Kennes, R., The transferable belief model. *Artificial Intelligence* **66**(2), 191-234.

[Snow, 1991] Snow, p., Compressed constraints in probabilistic logic and their revision. *Proceedings of the Seventh Conference on Uncertainty in Artificial Intelligence.* D'Ambrosio, B.D., Smets, Ph., and Bonissone, P. P. (eds), Morgan Kaufmann, San Mateo, 386-391.

[Spiegelhalter, 1986] Spiegelhalter, D.J., A statistical view of uncertainty in expert systems. *Artificial Intelligence and Statistics*, Chapter 2. Gale, W.A. (ed.), Addison-Wesley (also available in *Readings in Uncertainty Reasoning*, Chapter 5: 313-331, Shafer, G. and Pearl, J. (eds.), Morgan Kaufmann, San Mateo).

[Spohn, 1990] Spohn, W., A general non-probabilistic theory of inductive reasoning. *Proceedings of the Sixth Conference on Uncertainty in Artificial Intelligence.* Shachter, R., Levitt, T., Kanal, L.N., and Lemmer, J.(eds), Elsevier, New York, 149-158.

[Stephanou and Sage, 1986] Stephanou, H.E. and Sage, A.P., Perspectives on imperfect information processing. *IEEE Trans. Syst., Man, and Cybern.* **17**(5), 780-798.

[Tessem, 1993] Tessem, B., Approximations for efficient computation in the theory of evidence. *Artificial Intelligence* **61**, 315-329.

[Tsumoto and Tanaka, 1993] Tsumoto, S., and Tanaka, H., PRIMEROSE: Probabilistic rule induction method based on rough set theory. *Proceedings of Rough Sets, Fuzzy Sets and Knowledge Discovery.* Ziarko, W. (ed.), Springer-Verlag, Berlin, 274-281.

[Voorbraak, 1988] Voorbraak, F., A computational efficient approximation of Dempster-Shafer theory. *Int. J. Man-Machine Studies* **30**, 522-236.

[Voorbraak, 1991] Voorbraak, F., On the justification of Dempster's rule of combination. *Artificial Intelligence* **48**, 171-197.

[Wang *et al.* 1996] Wang, L., Wong, S. and Yao, Y., On the completeness of incidence calculus. *Int. J. Automated Reasoning* **16**, 355-358.

[Wellman, 1990a] Wellman, M., *Formulation of Tradeoffs in Planning under Uncertainty.* Pitman, London.

[Wellman, 1990b] Wellman, M., Qualitative Probabilistic Networks for Planning under Uncertainty. In *Readings in Uncertain Reasoning.* Shafer and Pearl (eds.), Morgan Kaufmann, San Mateo, 711-722.

[Wellman, 1993] Wellman, M., Henrion, M., Explaining 'explaining away'. *IEEE Trans. on Pattern Analysis and Machine Intelligence* **15**(3), 287-291.

[Wesley, 1983] Wesley, L.P. Reasoning about control: the investigation of an evidential approach. *Proceedings of the 8th International Joint Conference on Artificial Intelligence.* Bundy, A. (ed.), Karlsruhe, FRG, August, William Kaufmann, 203-206.

[Wells, *et al*, 1998a] Well, N., Liu, W., and Adamson, K., Using the ATMS for fault management in telecommunication networks. *Proceedings of Seventh Int. Conference on Information Processing and Management of Uncertainty in Knowledge-Based Systems (IPMU'98)*, Paris, 1816-1823.

[Wells, *et al*, 1998b] Wells, N., Liu, W., and Adamson, K., Using the ATMS for Telecommunication Network Fault Management. *Proceedings of Digest of FastAbstracts: the 28the Annual International Sysposium on Fault-Tolerant Computing* June 23-25, Munich, Germany, 31-32.

[Wilson, 1991] Wilson, N., A Monte-Carlo algorithm for Dempster-Shafer belief. em Proceedings of the Seventh Conference on Uncertainty in Artificial Intelligence. D'Ambrosio, B.D., Smets, Ph. and Bonissone, P.P. (eds.), Morgan Kaufmann, San Mateo, 414-417.

[Wong et al, 1992] Wong, S.K.M., Wang, L.S. and Yao, Y.Y., Interval structure: A framework for representing uncertain information. *Proceedings of the Eighth conference on Uncertainty in Artificial Intelligence.* Dubois, D., Wellman, P., D'Ambrosio, B. and Smets, Ph. (eds.), Morgan Kaufmann, San Mateo, 336-343.

[Wong and Wang, 1993] Wong, S.K.M. and Wang, Z.W., Qualitative measures of ambiguity. *Proceedings of the Nineth Conference on Uncertainty in Artificial Intelligence.* Heckerman, D. and Mamdani, E.H. (eds.), Morgan Kaufmann, San Mateo, 443-452.

[Wong and Ziarko, 1987] Wong, S.K.M., and Ziarko, W., Comparison of the probabilistic approximate calssification and the fuzzy set model. *Fuzzy Sets and Systems* **21**, 357-362.

[Wong, 1995] Wong, S.K.M., Interval structure: A qualitative measure of uncertainty. *Soft Computing: Rough Sets, Fuzzy Logic, Neural Networks, Uncertainty Management, Knowledge Discovery.* Lin and Wildberger (eds.), Sponsored by the Society for Computer Simulation, 22-27.

[Wong and Nie 1993] Wong, S.K.M. and Nie, X., Rough sets: a special case of Interval structures. *Proceedings of the International Workshop on Rough Sets and Knowledge Discovery (RSDK'93).* Ziarko (ed.), Springer-Verlag, 217-226.

[Wong, 1998] Wong, S.K.M., A rough-set method for reasoning about knowledge. In *Rough Sets and Knowledge Discovery 1.* Polkowski, L. and Skowron, A. (eds.). *Sudies in Fuzziness and Soft Computing* Vol 18, Physica-Verlag (Springer-Verlag), Heidelberg and New York, 276-285.

[Wong and Hwang, 1993] Wong, Y.C. and Hwang, S.Y., On parallelizing the Dempster-Shafer method using transputer network. *Parallel Computing* **19**, 807-822.

[Xiang *et al.*, 1993] Xiang, Y, Wong, S.K.M. and Cercone, N., Quantifying uncertainty of knowledge discovered from databases. *Proceedings of Rough Sets, Fuzzy Sets and Knowledge Discovery.* Ziarko, W. (ed.), Springer-Verlag, 63-73.

[Xu, 1995] Xu, H., Computing marginals for arbitrary subsets freom marginal representation in Markov trees. *Artificial Intelligence* bf 74, 177-189 (Research Note).

[Yager, Fadrizzi, Kacprzyk, 1994] Yager, R.R., Fedrizzi, M. and Kacprzyk, J. (eds). *Advances in the Dempster-Shafer Theory of Evidence* Wiley, New York.

[Yao and Li, 1993] Yao, Y.Y. and Li, X., Uncertain reasoning with interval-set algebra. *Proceedings of the International Workshop on Rough Sets and Knowledge Discovery (RSDK'93).* Ziarko, W. (ed.), Springer-Verlag, 178-185.

[Yao and Wong, 1992] Yao, Y.Y. and Wong, S.K.M., Representation, propagation and combination of uncertainty information. *Int. J. General Systems* **23**, 59-83.

[Yen, 1989] Yen, J., Gertis: A Dempster-Shafer Approach to Diagnosing Hierarchical Hypotheses. *Comm. ACM* (May), 573-578.

[Zadeh, 1975] Zadeh, L.A., Fuzzy logic and approximate reasoning. *Synthese* **30**, 407-428.

[Zadeh, 1984] Zadeh, L.A., A mathematical theory of evidence (book review). *AI Magazine* **5**(3), 81-83.

[Zadeh, 1986] Zadeh, L.A., A simple view of the Dempster-Shafer theory of evidence and its implication for the rule of combination. *AI Magazine* **7**, 85-90.

[Ziarko, 1993a] Ziarko, W.P., Variable precision rough set model. *J. Computer and System Sciences* **46**, 39-59.

[Ziarko, 1993b] Ziarko, W.P. (ed.), *Rough Sets, Fuzzy Sets and Knowledge Discovery, Proceedings of the International Workshop on Rough Sets and Knowledge Discovery.* Springer-Verlag.

Index

Mathematical Notation

P:	a set of atomic propositions
q or q_j:	an atomic proposition
$\mathcal{A}t$:	the basic element set formed from P
$\mathcal{L}(P)$:	the language set formed from P
ϕ, ψ:	formulae in propositional logic
$\psi \models \phi$:	formula $\psi \rightarrow \phi$ is valid (a tautology)
$\psi = \phi$:	when $\psi \models \phi$ and $\phi \models \psi$
$\mathcal{A}$:	a set of axioms
δ:	a basic element
X or S:	a set or space
χ:	a σ-algebra of a set X
χ' :	basis of χ
$\mathcal{W}$:	a set of possible worlds
μ, p:	a probability distribution on a set
i:	an incidence function
i_*, inf:	lower bound of incidences
i^*, sup:	upper bound of incidences
ii:	a basic incidence assignment
G:	an assignment with a pair of inf_G and sup_G
$Prob_*$:	lower-bound of a probability distribution
$Prob^*$:	upper-bound of a probability distribution
Θ:	a frame of discernment
m:	a mass function on a frame
bel:	a belief function on a frame
pls:	a plausibility function on a frame
A_{DS}:	a set containing all the focal elements of a belief function
Γ:	a multivalued mapping function
$\otimes$:	set product. $X_1 \otimes X_2 = \{< x_{1i}, x_{2j} > \mid x_{1i} \in X_1, x_{2j} \in X_2\}$
U:	a universe
$\emptyset$:	the empty set
Π:	possibility measure
N:	necessity measure
R:	an equivalence relation

$\Rightarrow$:	$a, b \Rightarrow c$ means if the antecedents are true, then the consequent is true
$\Leftrightarrow$:	$a, b \Leftrightarrow c$ means $a, b \Rightarrow c$ and $c \Rightarrow a, b$

List of Figures

List of Tables

www.ingramcontent.com/pod-product-compliance
Ingram Content Group UK Ltd.
Pitfield, Milton Keynes, MK11 3LW, UK
UKHW021933200726
13853UKWH00010B/1285

* 9 7 8 3 6 6 2 0 0 3 4 4 2 *